5448

Transboundary Water Resources: Strategies for Regional Security and Ecological Stability

NATO Science Series

A Series presenting the results of scientific meetings supported under the NATO Science Programme.

The Series is published by IOS Press, Amsterdam, and Springer in conjunction with the NATO Scientific Affairs Division

Sub-Series

I. Life and Behavioural Sciences	IOS Press
II. Mathematics, Physics and Chemistry	Springer
III. Computer and Systems Science	IOS Press
IV. Earth and Environmental Sciences	Springer
V. Science and Technology Policy	IOS Press

The NATO Science Series continues the series of books published formerly as the NATO ASI Series.

The NATO Science Programme offers support for collaboration in civil science between scientists of countries of the Euro-Atlantic Partnership Council. The types of scientific meeting generally supported are "Advanced Study Institutes" and "Advanced Research Workshops", although other types of meeting are supported from time to time. The NATO Science Series collects together the results of these meetings. The meetings are co-organized bij scientists from NATO countries and scientists from NATO's Partner countries – countries of the CIS and Central and Eastern Europe.

Advanced Study Institutes are high-level tutorial courses offering in-depth study of latest advances in a field.
Advanced Research Workshops are expert meetings aimed at critical assessment of a field, and identification of directions for future action.

As a consequence of the restructuring of the NATO Science Programme in 1999, the NATO Science Series has been re-organised and there are currently five sub-series as noted above. Please consult the following web sites for information on previous volumes published in the Series, as well as details of earlier sub-series.

http://www.nato.int/science
http://www.springeronline.com
http://www.iospress.nl
http://www.wtv-books.de/nato-pco.htm

Series IV: Earth and Environmental Sciences – Vol. 46

Transboundary Water Resources: Strategies for Regional Security and Ecological Stability

edited by

Hartmut Vogtmann
Federal Agency for Nature Conservation,
Bonn, Germany

and

Nikolai Dobretsov
Siberian Branch of Russian Academy of Sciences,
Novosibirsk, Russia

in collaboration with

Astrid Mittelstädt

Springer

Proceedings of the NATO Advanced Research Workshop on
Transboundary Water Resources: Strategies for Regional Security and Ecological Stability
Novosibirsk, Russia
25–27 August, 2003

A C.I.P. Catalogue record for this book is available from the Library of Congress.

ISBN 1-4020-3080-0 (HB)
ISBN 1-4020-3082-7 (e-book)

Published by Springer,
P.O. Box 17, 3300 AA Dordrecht, The Netherlands.

Sold and distributed in North, Central and South America
by Springer,
101 Philip Drive, Norwell, MA 02061, U.S.A.

In all other countries, sold and distributed
by Springer,
P.O. Box 322, 3300 AH Dordrecht, The Netherlands.

Printed on acid-free paper

Printed in the Netherlands.

TABLE OF CONTENTS

Chapter 4: Ecological and economic aspects of water management

Chapter 5: Scientific analysis and tools of water changes

Chapter 6: Strategic implications of water access

PREFACE

After the soviet era and since their independence the new Central Asian countries are rebuilding a system of water resources management: an important challenge for the development of the whole region. The NATO workshop held on 25-27 August 2003 by the Federal Agency for Nature Conservation, Germany and the Siberian Branch of the Russian Academy of Science, attended by experts from five Central Asian countries, Russia, six Western European countries, the US and the UNEP offered water engineers and nature scientists as well as economic and political scientists and practitioners from water administrations and international river commissions to meet in Novosibirsk and develop sustainable approaches in the management of Central Asian water resources.

This book presents important aspects of transboundary water resources, i.e. the global water crisis: problems and perspectives; regional experiences in solving water problems in Central Asia; problems and management of transboundary water resources; ecological and economic aspects of water management; scientific analysis and tools of water changes; strategic implications of water access arisen during the workshop.

A final recommendation in the area of equitable sharing of benefits, monitoring and data collection as well as proposals for Central Asia transboundary waters programme were set in the book as the main result of the meeting.

The editors.

ENVIRONMENTALLY SUSTAINABLE WATER USE FOR SUSTAINABLE DEVELOPMENT AND ENHANCING SECURITY IN CENTRAL ASIA

SMITH, D.H.
Economist, Division of Policy Development and Law, United Nations Environmental Progamme

ABSTRACT

This chapter outlines the relationship between environmentally sustainable water use, sustainable development and regional security in Central Asia. Environmentally sustainable freshwater use is argued to be an essential component of sustainable development, including poverty reduction. It is also argued to be important for reducing the potential for insecurity in Central Asia, in the context of transboundary waters in the region. The chapter sets out key elements for environmentally sustainable water use and successful transboundary waters management.

1. INTRODUCTION – THE GLOBAL SITUATION

Freshwater is a fundamental requirement for human survival and socio-economic development. Chapter 18 of Agenda 21 highlights the importance of water and indicates the way to a secure, sustainable water future. Its general objective is to ensure adequate supplies of water of good quality for the human population of the Earth, while also preserving the hydrological, biological and chemical functions of ecosystems and combating vectors of water-related diseases. Most importantly, however, Chapter 18 advocates that humans adapt their activities to live within the capacity limits of nature. That is, we must change the way we manage and use water to achieve the sustainable use of water. It is stressed that this is not primarily for the sake of the environment; rather it is to secure sustainable economic and social development for people, especially poor people.

Yet we are far from achieving the sustainable use of water and in many parts of the world, people are facing a water crisis. And, unlike the energy crisis, the water crisis is life threatening, and it is the most immediate and serious environmental, social and economic problem facing over a billion people in the world today.

The extent of the water crisis is indicated by the following examples (UNEP 1999; 2002):

- World-wide, about two thirds of the population in 2025 are likely to be subject to moderate to high levels of water stress
- About 20% of the world's population lacks access to safe drinking water and 50% lacks adequate sanitation.

H. Vogtmann and N. Dobretsov (eds.), Transboundary Water Resources: Strategies for Regional Security and Ecological Stability, 1–25.

- World-wide, three million people die every year from diarrhoeal diseases (such as cholera and dysentery) caused by contaminated water;
- Polluted water affects the health of 1.2 billion people every year, and contributes to the death of 15 million children under five every year.
- Vector-borne diseases, such as malaria, kill another 1.5 to 2.7 million people per year, with inadequate water management a key cause of such diseases.
- Increasing demands on available water supplies carry the potential for conflict over transboundary water resources.

At present levels of population and water demand, there is not enough clean water in large parts of the world. If water management remains as inadequate as it currently is, the present water crisis will become a catastrophe that will prevent the achievement of sustainable development in many parts of the world and cause the deaths of millions more people, mainly children. And while the crisis is not so life threatening in many parts of the world, water related economic costs and environmental degradation are also serious issues in many countries. Additionally, increased political tension as competition for water use intensifies in some water basins is more likely.

2. CONTRIBUTION OF THE ENVIRONMENTALLY SUSTAINABLE WATER USE TO SUSTAINABLE DEVELOPMENT

Water is a key element of sustainable development because it is an essential component of life and income generating activities. But for too long the environment has been seen as a competing user. A response to criticisms of environmentally unsustainable water use may be that there is no other choice but to use water in an environmentally unsustainable way. However, this view of the environment as a competitor misses the critical point that the environment is fundamental to sustainable development. That is, if water is used unsustainably over time less is available to meet the needs of people. More specifically, if water is used consistently at a faster rate that it is replenished and/or if it is polluted so its use is restricted, then there are direct economic and social costs. As the World Bank states "the environment [is] not just another consumptive user of water but the water resource itself and … degrading the quantity and quality of water in rivers, lakes, wetlands and aquifers can inextricably alter the water resources system and its associated biota, affecting present and future generations." (World Bank, 1993). (The discussion above follows Smith et al 1998).

The sustainable use of water provides sustainable economic benefits - through good health and income generating activities, including food production. Conversely, unsustainable water use imposes costs through health problems and reduced production. While there is much focus on the costs of environmental protection, there is not enough focus on the costs of using environmental resources such as water unsustainably. This is partly because these costs are often borne by those who have the least say - the poor.

Costs of environmentally unsustainable water use

Health Costs

The most obvious water-related cost is from the sickness and death some forms of pollution cause. The costs of sickness and death from water pollution are enormous, but because they are often borne by individuals with little or no market or political power, those in power and who manage water do not pay sufficient attention to these costs. Taking the figure quoted at the beginning of this paper of 1.2 billion people falling ill from contaminated water per year, if US$5 is needed per person for medicines, that is equivalent to US $ 6.0 billion per year in medical costs alone.

As the burden of such sickness and death falls on the poorer members of society, the cost as a proportion of their income is much higher than it would be in the developed world. If someone earns about a US$1.00 a day, and it costs US$5.00 to treat a water borne illness, a very heavy financial burden is imposed on that person.

Production and Income Costs

As indicated above, water generates benefits for humans. People use water to produce, for example, food, and generate revenue from sell what they produce. Thus, if water is used in an environmentally sustainable manner over time, the productive capacity of water resources is reduced and production and incomes decline.

The case of the Aral Sea basin is a well-known example of how environmentally unsustainable water use can have serious economic consequences. The environmentally unsustainable use of freshwater resources for agricultural irrigation in the Aral Sea basin has decreased lake water levels and quality severely. As a result, the fishing industry has collapsed. In addition, the inefficient use of irrigation water in this semi-arid region led to salinisation and a subsequent decrease in agricultural production. Serious human health problems have arisen with wind blown dust contaminated by agricultural chemical residues. (UNEP 1993; UNDP 1996).

Ecosystem costs

While pollution and excess water withdrawals cause serious damage to ecosystems in many parts of the world, which can be quantified in physical terms, it is much more difficult to calculate the monetary value of such damage. Declines in fish catches, for example, can be valued, as there is a direct link between ecosystems and human income. Less easily, health cost and productive impacts from pollution related illnesses can also be estimated. Biodiversity decline is far more difficult to value. Nevertheless, these impacts and cost occur, and sooner or later, reduce the goods and services that water provides to humans. Thus they must be included in a comprehensive cost-benefit analysis. But unfortunately these impacts are not adequately considered in many cases.

Thus, using water unsustainably wastes water and wastes money, both of which are essential elements of sustainable development. To reiterate, environmentally sustainable water use is a vital component of sustainable economic and social development.

And it is important to stress that sustainable water and land use are closely connected. Land in water catchments must be managed sustainable to maintain and maximise the productivity of water. For example, unsustainable land use causes erosion that results in sediment run-off to water bodies and reduced water retention, increased flooding and lower reservoir storage, as well as decreasing soil fertility.

3. CONTRIBUTION OF ENVIRONMENTALLY SUSTAINABLE WATER USE TOWARDS POVERTY REDUCTION

Not only is environmentally sustainable water use a pillar of sustainable development, but also coupled with appropriate management regimes, it could and should be a vital contributor to poverty reduction.

As indicated above, health costs arising from diseases caused by polluted water impose costs the poor are least able to bear. In addition, excess withdrawals and pollution reduce the economic benefits from water use. Conversely, sustainable water use and sound management produce sustainable income from the goods and services requiring water – most obviously food. Improvements in water resources management, if targeted at the poor, can have a major impact on their lives by enabling them to increase productive activities that result in significant increases in income, thus reducing poverty.

In terms of water supply and sanitation, sound water utility management generates sustainable revenue flows which can then be used to improve supply to poor people and help all water users use water more efficiently through investment in appropriate end use technologies. Water utility management inefficiencies contribute directly to poverty - in many cities water losses are 40% or more of total water supply! Conversely, water utility efficiency improvements would contribute directly to improving the environmental sustainability of water use and reducing poverty, with improved services reducing illness and associated costs, and increasing the potential for increases in income.

4. CONTRIBUTION OF THE ENVIRONMENTALLY SUSTAINABLE WATER USE TOWARDS REGIONAL SECURITY IN CENTRAL ASIA.

Increasing competition over water resources in Central Asia could lead to increased political tension, due to, *inter alia*, disagreements over the allocation of water and the sharing of associated benefits between states.

Environmentally unsustainable water use increases the potential for tensions to develop, because unsustainable use reduces the quantity and/or quality of water resources and as a result reduces the actual or potential benefits generated. Thus states are faced with a declining resource, and if demand for water is constant or increasing, competition for this declining resource will increase, thereby raising the potential for disputes and increased tensions.

Environmentally sustainable water use over time maintains the quality and quantity of the water available and therefore the benefits that can be generated from the water resources in question. Therefore the potential for disputes is reduced if such water use is maintained, compared with using water unsustainably. The potential for disputes can be reduced if water management improves the environmental sustainability of water use and increases the potential benefits that can be generated.

A number of factors –political, geographical, economic, social and environmental – combine to determine the state of regional security and it is necessary to fully include the environmental dimension in regional security considerations.

Given the transboundary nature of water resources in Central Asia, it is important that a transboundary approach is taken to manage such waters, to maximise the benefits generated by the water resources and to share them equitably to, *inter alia*, reduce the potential for regional disputes.

In summary, environmentally sustainable water use managed on a transboundary basis is consistent with reducing the potential for regional insecurity. Below key elements for environmentally sustainable water management are set out, followed by a section on transboundary water management.

5. ACHIEVING ENVIRONMENTALLY SUSTAINABLE WATER USE: SOME KEY ELEMENTS.

The tools and techniques necessary for achieving environmentally sustainable water use have been agreed upon by the international community – as evidenced by the Chapter 18 of Agenda 21, (1992) the Bonn International Conference on Freshwater (2001) and the World Summit on Sustainable Development. (2002). The prime issue is to ensure the implementation of agreed targets using generally accepted and recognised techniques and tools.

It is also necessary to ensure the actions and policies in key water using sectors are consistent with environmentally sustainable water use. Thus, an intersectoral approach is needed.

Governance

Good governance is the most important perquisite to solving freshwater problems. Governance includes policy, legal frameworks, management approaches, institutional structures and decision-making processes. Another essential requirement for good governance is the political will to address the problems, including corruption and inefficiency.

This is widely recognised as the key reason why water is used in an environmentally unsustainable manner and why so many people lack access to clean water and waste treatment infrastructure. This recognition is long standing. For example, the Conference on Water Development in Less Developed Areas, held in Berlin in 1963, found that "the main problems in the field are not technical, but are of an organisational, administrative, political or managerial nature Aid giving nations may be more helpful in solving these political problems than in providing either engineering or financial aid." (Howe, 1995).

Despite this recognition, attempts to address water management and policy issues have been grossly inadequate. The international community and national governments share responsibility for this. It is easier to provide money for infrastructure projects than train managers and reform water management agencies, which is probably partly why donors in the past have concentrated on such projects. To date the focus has been too much on engineering works - and many of these have failed to deliver anticipated benefits because the management frameworks in which they were planned and build were inadequate. Sound engineering works flow out of sound management and planning frameworks – that is, out of sound governance.

For example, around Lake Chad are hundreds of metres of unusable concrete irrigation channels and associated pumps. These are unusable because unsustainable water use from the lake has reduced water levels to below the level that can be reached by the pumps.

Below are some of the key elements needed for improving water governance.

Water Policy and Legal Frameworks

Water policies and law should:

- Have an overall objective of environmentally sustainable water use. Environmentally sustainable water use is the key to maximising, over time, the sustainable benefits from water resources.
- Incorporate this objective in policy detail so that it is operationally meaningful

It is not enough to have an overall policy objective of environmentally sustainable water use. The objective must be reflected in the operational clauses of water policy and law detail so that it results in the environmentally sustainable use of water resources

- Mandate the use of Integrated Water Resources Management (IWRM)

IWRM is the key tool to environmentally sustainable management of water and maximising the benefits of water resources use over time. (See below).

- Avoid fragmented water use & allocation decision-making and ensure there is inter-sectoral policy co-ordination

Co-ordinated decision-making on water use is vital for ensuring that water goes to the highest value uses, and thus generates the maximum benefits for people. It is necessary to ensure the actions and policies in key water using sectors are consistent with sustainable water use. Thus, an intersectoral approach is needed. For example, it does not make sense to invest significant capital in additional irrigation schemes when cheaper improvements in food storage to cut food spoilage would meet food needs. If agricultural policy and pricing encourage inefficient use of water, these should be addressed. As another example, industries should not be permitted to pollute water and profit from this, especially when the poor have to use the polluted water for their basic needs.

- Include enforceable incentives

Enforceable incentives – both economic and regulatory – are vital to ensuring water use is efficient and environmentally sustainable. Enforced regulations and fines applying to pollution plus economic incentives for efficient water use needed. (Meister, 1995).

- Adopt the subsidiarity principle

This is that water use decision-making should be devolved to the lowest government level consistent with internalising significant externalities. For example, management of a village water hole should be devolved to the village, within national policy and legal guidelines. And on the other end of the scale, management of a transboundary water resource should be conducted on an inter-governmental basis.

Appropriate Institutional Frameworks

Appropriate institutional structures and mandates are vital for good governance and integrated water resources management. Various models for appropriate water management institutions exist, and different models have been proven to work. Whatever the final design of water management institutions, there are certain goals and characteristics that they should aim for. From a 1995 UNEP workshop on economic principles from water management (Howe, 1995):

The institutional framework should ensure:

- Co-ordinated surface water and groundwater management;
- Co-ordinated water quantity and water quality management;
- The provision of incentives for greater economic and physical efficiencies in water use;
- Protection of in-stream flow values and other public values related to water systems.

This requires water management institutions that have the following characteristics (Howe, 1995):

i. Capability of co-ordinating water plans and management procedures with other functional agencies. (E.g. agriculture, environment, economic planning, industry.

ii. Capability of considering a wide range of alternative solutions to water problems, including non-structural measures and the use of economic instruments. (E.g. pricing, taxes, tradable permits, subsidies, etc.).

iii. Separate planning and evaluation from construction and management functions. (I.e. Do not have dam building agencies responsible for watershed management.)

iv. Have the multi-disciplinary expertise to carry out "multiple-objective planning and evaluation." (See Multi-objective planning below.)

v. Observe the "subsidiarity principle" in assigning responsibilities to agencies at national, provincial and local levels. (IE assign responsibilities to the lowest level consist with the internalisation of important externalities.

vi. Have the expertise to involve all "stakeholders" in planning operations from the beginning.

vii. Build in a reward structure that will stimulate creativity and innovation.

viii. Build in a reward structure that will stimulate learning through ex post analyses.

Multi-Objective Planning

Appropriate tools must also be used. A key tool for integrated, inter-sectoral approaches is multi-objective planning that bases catchment management decisions on an integrated assessment of environmental, economic and social factors. (Howe, 1995). As noted above, environmental degradation inevitably has economic and social consequences for human beings. In the case of catchment degradation, agriculture and energy production are reduced, which imposes direct economic costs. Contaminated water imposes harsh economic and social costs on people.

The elements of multi-objective planning are:

i. Cost-benefit analysis from the national perspective
ii. Cost-benefit analysis from the project or regional perspective
iii. Environmental impact analysis
iv. Social impact analysis (usually non-monetised.)

Cost-benefit analysis should include economic, social and environmental impacts. If it does not, then it will not include all the relevant impacts. Some impacts cannot be costed, but they should at least be identified. A diagram on how to integrate environmental, economic and social assessment is in Annex1. "A narrow benefit-cost analysis would include only those factors outlined in double lines by extending the benefit-cost analysis, as indicated by the single lines, the whole array of effects on the natural system, the receptors, and the economy are incorporated." (Hufschmidt et al, 1983).

Including the poor in water use decisions.

An enhanced partnership between the poor majority, (who lack adequate access to water and sanitation), government and private sector partners is vital. Currently those who have access to water are far more prominent in decisions about water supply than those who do not have water. Thus, the interests of existing users predominate over the interests of the poor. No matter that the needs of the poor are desperate, it is too often the case than present users have more impact on water use investment and management decisions. Therefore it is important that those currently without adequate supplies of water are provided with formal and informal mechanisms so they have an active say in water policy and infrastructure decision-making.

Integrated Water Resources Management. (IWRM).

IWRM is the accepted key tool for successful water resources management. A core principle of IWRM is the environmentally sustainable use of water. Integration occurs at all levels and for all aspects. For example, IWRM includes integrated social, economic and environmental management, integrated sectoral management, plus integrated land- water management. It is strongly recommended that it is applied on a water catchment basis, as the water catchment is the natural management unit, encompassing all significant interactions in a water body.

There are a number of detailed guidelines on IWRM – for example, see the Global Water Partnership IWRM Toolbox (http://www.gwpforum.org/servlet/PSP).

Mobilising Financial Resources for water resources management and service provision.

Adequate financing is vital to achieve environmentally sustainable water management and to expand service provision, and good governance is vital to raising finance. But mobilising finance can be a highly sensitive issue, because it is includes such issues as privatisation and water pricing. Finance mobilisation solutions that benefit developing countries in the long term, address the needs of the poor majority and that are consistent with the environmentally sustainable management of water are the aim.

There is no doubt that a significant increase in international funding for water sector reform and infrastructure is vital and urgent. Billions more are needed, but they must be spent wisely.

With regard to privatisation, the sale of water utilities and the contracting out of water revenue collection have advantages - such as service delivery improvements, improved revenue collection and financing for expansion. But there are also disadvantages. If developing country utilities are sold in a poor management and financial state then the price gained is lower than if management was efficient and the financial situation good. In addition, the revenue stream to government or local authorities from water sales is reduced or lost. Privatised utilities require regulation, and in the developed world sophisticated regulatory authorities have been established to monitor privatised water utilities. However, most developing countries do not have the capacity to establish these.

A suggested approach is for the international community and governments to place far more emphasis on improving the management of developing country water utilities. Then improved revenue streams will be generated to finance service delivery and infrastructure improvements - and also to finance improvements to enhance the environmental sustainability of water supply and use. This could be achieved through donor funded management contracts where donors and private sector experts enhance the capacity of water utilities to manage themselves.

Water Pricing

Water pricing is a key tool for efficient management of water resources. Water sales are the most direct way of raising finance and appropriately applied, they also encourage improved efficiency in water use. There is no doubt that over time, water prices that reflect the true cost of water supply will result in more efficient use of water and more cost effective investment in water infrastructure. But the needs of the poor majority must be met at a price they can afford.

While some reject the use of pricing that reflects the cost of supply, consideration should be given to the costs of subsidising water to larger users. The opportunity cost to poor people is one cost - water used inefficiently by larger users and the funds used in infrastructure to supply such inefficient use could be used in supplying the needs of the poor. As it is, in many places the poor pay more per litre of water than the rich. For example, in Nairobi, water vendors in the slums often charge about 25 US cents per twenty litres. In contrast, in the better suburbs of Nairobi, water prices range from about 35 US cents to 70 US cents *per cubic metre* of water! And the costs the water for poorer people all too often include sickness.

Therefore it is not surprising that when the poor are asked, research suggests (Sivalingnam 1995, 1.) that they have a higher willingness-to-pay for clean, secure water supplies than richer people tend to assume. Moreover, in many cases, the cost of piped water is lower than the poor currently pay.

For larger users, the questions must be asked: If they are making money from selling goods why cannot they pay the true cost of water? What proportion of their production costs is comprised of water? Do they have options to improve the efficiency with which they use water, and thus reduce their costs if water prices rise? (Total cost equals price times quantity - if the price goes up, but the quantity used goes down through more efficient use, total costs can remain relatively stable).

It is important to stress that the long-term costs of providing subsidies to the non-poor by failing to reflect the value of water in the price of water and waste disposal into water are very high and fall mainly on the poor. These costs are worth considering in some depth.

The costs of subsidies, explicit and hidden

Explicit Subsidies

Large explicit price subsidies are granted in many countries to large-scale water users, who pay significantly less than the cost of supplying water. When water is subsidised it tends to be wasted. For example, in large parts of Southern Africa, 80% of all water consumption is in agriculture - and about 65% of this is wasted! This part of the world is facing serious water shortages and such waste is increasingly costly. Thus water price subsidies impose a direct cost on other taxpayers and also an indirect cost through encouraging inefficient water use.

Explicit subsidies also have an opportunity cost. The funds used for subsidising water might be better used by society for other purposes, examples being to pay for water supply to other areas or for health clinics. Moreover, the inefficient use of irrigation water has major environmental impacts - including salinisation, which means agricultural production is severely reduced or even halted in some areas.

Although water subsidies may be justified in some situations, such as helping the poor access a basic level of supply, it is important that decision makers calculate all the significant costs and impacts of providing subsidies for water resources before a decision is made to apply them.

Subsidies in other sectors

Subsidies in other sectors can have the same impact as direct water price subsidies, creating incentives for unsustainable water use. Agricultural and energy subsidies encourage activities that result in inefficient water use and increased pollution. For example, production subsidies in agriculture encourage higher consumption of water, fertilisers and pesticides at a high cost to the taxpayer. This increases pressure on water resources and pollutes land and water. This in turn can increase demands for expenditure on remedial environmental protection and taxpayers are faced with a double bill as a result. (One for agricultural subsidies, another for action to clean up the environmental damage encouraged by agricultural subsidies).

Hidden subsidies

An example of a hidden subsidy is when people pollute water and do not pay. This transfers the costs of pollution to other users, and in some cases these costs are enormous. For example, in the case of pollution that results in illness and death or destroys fish stocks.

Sector Level Resource Misallocation Costs

These costs are rarely considered. Where unsustainable use of water occurs due to sector wide policy - such as agricultural subsidies and subsidies for irrigation projects and irrigation water - there can be a large-scale misallocation of resources. That is, a much higher level of investment in irrigation and irrigation-based activities occurs than would have been the case if full costs were paid for irrigation projects and water. This is a misallocation when funds invested in irrigation could have been used elsewhere in the economy for higher value uses - such as health clinics or growing different types of food or reforestation or in domestic or industrial uses. The large number of criticisms and negative evaluations of water projects suggests that billions of dollars have been wasted on large-scale water infrastructure projects.

However, the concerns about the impacts of water price increases are perfectly valid. Below is an approach designed to cater for such concerns.

Progressive Pricing - A practical approach that helps the poor and provides economic incentives for environmentally sustainable and economically efficient water use.

There are different models for pricing water – with examples including declining block tariffs, marginal cost pricing and progressive pricing. (Sivalingnam, 1995; 2.). Progressive pricing is the preferred model from efficiency, equity and environmental sustainability grounds. It involves setting a price structure that charges more per unit the more water is used. A basic needs amount of water should be sold at a low price, subsidised if necessary, such that poor people can afford the minimum needed for a healthy existence. Increasing levels of consumption are charged for at progressively higher tariffs per unit sold. There is sound economic justification for such an approach. Supplying a basic amount of water to each person requires a lower level of investment in water infrastructure than does a higher level of supply to users. In addition, the public good benefits of water subsidies for the poor and likely to be high.

While existing large-scale users may protest that they cannot afford to pay unsubsidised prices for water, careful analysis may suggest this is not true over the long term. It is recommended that some of the higher revenue from water price increases be used to assist larger users to improve the efficiency with which they use water - for example through investment in more modern irrigation technologies. But this assistance may also be economically efficient, in that water saved through current users adopting more efficient technologies can be used elsewhere, and the water saved at lower cost than that of building a new reservoir.

In a more general sense, fixing leaks and largely eliminating water theft can have a dramatic impact on a water utilities cash-flow and reduce the prices that need to be charged to cover costs.

The secure, permanent increase in water utility revenue streams enables an expansion in water supply. Such a secure cash flow can be used to secure loans for improvements in service delivery.

Pollution Taxes

With regard to including the cost of pollution in water charges, the issue of pollution taxes is also difficult and sensitive. In addition, such taxes may not have the desired effect unless they are set at unacceptably high levels. (Due to, for example, the impact of sunk capital costs). In the case of pollution, a combination of standards and financial incentives, both positive and negative, based on the impact on receiving waters is probably the most viable option. But some level of pollution tax is recommended as a disincentive to polluters but also a way of raising revenue to improve environmental quality.

Ensure women are a central part of water management

The international community has recognised – for example in Agenda 21 - that women play a central part in the provision, management & safeguarding of water – that is, in water governance. Women and children are the ones who suffer most seriously and directly from the absence of clean water - children are the most vulnerable and women their prime carers. Women are also responsible for the vast majority of the emotional and physical support the family needs to survive. Thus they have the most vital interest in water supply and sanitation issues. Experience shows that given appropriate training they are also excellent local water supply managers. Thus, international and national efforts should increase their focus on enabling women to participate more in water supply decisions and in the management of local water supply facilities - such as community taps, wells and other supply infrastructure.

Least-cost planning.

Least-cost planning is a seriously under-utilised water management tool. It involves considering both demand and supply side options for water supply augmentation. For example, including demand and supply side efficiency improvements in the list of possible water augmentation options. These efficiency improvements include fixing leaks, using more efficient irrigation technologies and rainwater harvesting. Saving water in one place means it can be used elsewhere. The basis of choice between options is cost – if it is cheaper to fix leaks than build a new reservoir, then a leak-fixing programme would be undertaken. Given that water loss rates in major urban areas in developing countries exceed 40% in a number of cases, this could save the construction of major reservoirs, which also have significant environmental impacts. Environmental impacts can be included in water supply option ranking through a cost weighting system – for example, placing a ten percent cost weighting on environmentally unsound options.

Other elements for environmentally sustainable water use besides Governance.

Other elements include capacity development and technology transfer.

Capacity Development

Given that governance issues have been recognised as the key to addressing freshwater problems, international and national efforts should focus on capacity development for management, institutional and governance reform. Not least because water supply is a revenue generating operation and sound management will improve revenue streams and thus enable improvements in supply, including engineering works.

Technology Transfer

The transfer of appropriate technology is vital. Many advances in technology enable more benefits to be generated per unit of water. If governance is appropriate, it is more likely that technologically advance options for water supply and use will be invested in. For example, water prices that reflect the cost of supply will encourage investment in more efficient technologies than subsidised water. Roof rainwater collection, efficient irrigation technology and leak location and repair technologies are examples of technology that needs to be invested in far more than is currently the case.

6. TRANSBOUNDARY WATERS MANAGEMENT: KEY ELEMENTS

Governance is a particularly important and complex issue with respect to transboundary waters. The implementation of international agreements, including the UN Convention of the Non-Navigational Uses of International Water Courses, is important, followed by more specific regional and water basin agreements. Real political commitment from all riparian states to act in a spirit of give and take and consider such issues as water and benefit sharing is the critical prerequisite to success in transboundary waters governance. Transboundary waters management can be described as IWRM plus diplomacy.

Key elements of successful transboundary waters management include:

Commitment from all riparian states to fairly address all issues, including water allocation and benefit sharing.

This is essential, for without it, at best, only partial planning and implementation of measures to maximise the benefits from the basin in an equitable and sustainable manner is likely.

Adoption of legal frameworks e.g. shared waters protocols & UN Convention on Non-Navigational Uses of International Waters.

Legal frameworks set the parameters for successful management, including dispute resolution. While the UN Convention sets important parameters and ratification is urged, more target sub-regional and water basin protocols are required. Dispute consideration and resolution mechanisms must be included. An example is the Protocol On Shared Watercourse Systems In The Southern African Development Community (SADC). (Annex 2).

Establishment of water basin governance & institutional structures with secure funding e.g. basin commissions.

These are essential for developing and implementing programmes to for the management of the river basin as a whole.

Application of IWRM on a water basin basis.

IWRM is the key tool to for maximising the benefits from water basins in an environmentally sustainable manner. Essentially, it can be described as the application of an integrated social-economic-environmental approach utilising multi-objective planning and integrated land-water planning.

Benefit sharing mechanisms

Maximising the economic development opportunities from a specific water basin does not necessarily mean dividing water resources equitably between states. It may mean using water more in one state where this may generate higher benefits overall. However, for this principle to be accepted, a mechanism for compensation, or benefit sharing, is required. The basic approach is to develop a plan to maximise overall basin benefits and then negotiate benefit sharing.

Development of a comprehensive transboundary water basin management programme & associated projects

To realise the potential economic benefits in a water basin - and in an environmentally sustainable manner - requires an operational programme and projects designed and implemented on a transboundary water basin basis. This could include a transboundary strategic action plan for the basin plus subsidiary projects. (For example, on capacity building, economic development, water and sanitation, water quality etc). These projects should be prioritised and implemented in an integrated manner. As an example, SADC adopted a Water Sector Regional Strategic Action Plan and then developed 31 priority projects to implement the plan, including on economic accounting for water, water policy and legal reform, capacity building, and on sanitation infrastructure development.

Policy & legislation harmonisation steps among riparian states

Management of a transboundary water basin will partly depend on the water policy and laws in each riparian state; therefore it is important that these policies and laws are in harmony. That does not mean that identical laws and policies are required. It does mean that, at least, inconsistencies that hamper joint management of a transboundary basin to maximise and share benefits need to be removed. For example, if an upstream country has pollution regulations that permit significant pollution to impact on a downstream country, then these should be reviewed.

7. CONCLUSION

Environmentally unsustainable freshwater use is a threat to economic development – including poverty reduction – and the environment. Using water faster than it is replenished and polluting it cuts incomes and increases costs and worsens poverty. It also threatens regional security in the context of transboundary waters in Central Asia, as environmentally unsustainable water use increases competition between states over water resources. Thus to reduce competition over water resources and reduce the potential for regional insecurity, environmentally sustainable water use is necessary.

8. ANNEX : PROTOCOL ON SHARED WATERCOURSE SYSTEMS IN THE SOUTHERN AFRICAN DEVELOPMENT COMMUNITY (SADC) REGION

Table of Contents

Protocol on shared watercourse systems in the Southern African Development Community (SADC) Region

Preamble

The Republic of Angola, the Republic of Botswana, the Kingdom of Lesotho, the Republic of Malawi, the Republic of Mozambique, the Republic of Namibia, the Republic of South Africa, the Kingdom of Swaziland, the United Republic of Tanzania, the Republic of Zambia and the Republic of Zimbabwe;

BEARING in mind the Helsinki Rules on uses of the waters of International Rivers and the work of the International Law Commission on the non-navigational uses of international watercourses;

RECOGNISING the relevant provisions of Agenda 21 of the United Nations Conference on Environment and Development, the concepts of environmentally sound management, sustainable development and equitable utilisation of shared watercourse systems in the SADC Region;

CONSIDERING the existing and emerging socio-economic development programmes in the SADC region and their impact on the environment;

DESIROUS of developing close cooperation for judicious and coordinated utilisation of the resources of the shared watercourse systems in the SADC region;

CONVINCED of the need for coordinated and environmentally sound development of the resources of shared watercourse systems in the SADC region in order to support sustainable socio-economic development;

RECOGNISING that there are as yet no regional conventions regulating common utilisation and management of the resources of shared watercourse systems in the SADC region;

MINDFUL of the existence of other Agreements in the SADC region regarding the Common utilisation of certain watercourses; and

IN ACCORDANCE with Article 22 of the Treaty establishing SADC, have agreed as follows:

Article 1

Interpretation of Terms

1. For the purposes of this Protocol the following terms shall have the meanings ascribed to them hereunder:
"Agricultural use" means use of water for irrigation purposes.

"Basin" means drainage basin.

"Basin State" means a State part or all of whose territory is within a drainage basin.

"Domestic Use" means use of water for drinking, washing, cooking, bathing, sanitation and stock watering purposes.

"Drainage Basin" means a geographical area determined by the watershed limits of a system of waters including underground waters flowing into a common terminus.

"Emergency situation" means a situation that causes, or poses an imminent threat of causing serious harm to Basin States and which results suddenly from natural causes, such as floods, landslides or earthquakes or from human conduct.

"Industrial use" means use of water for commercial, electrical power generation, industrial, manufacturing and mining purposes

"Member State" means a State which is a member of SADC, party to this protocol.

"Navigational use" means use of water for sailing whether it is for transport, fishing, recreation or tourism.

"Riparian Land" means land contiguous to, abutting on or overlying waters of a stream, lake or aquifer or land through which a watercourse passes.

"Riparian State" means a State through whose territory or along whose border a watercourse passes.

"Shared watercourse system" means a watercourse system passing through or forming the border between two or more basin states

"Watercourse State" means a State in whose territory part of a watercourse system is situated.

"Watercourse system" means the inter-related hydrologic components of a drainage basin such as streams, rivers, lakes, canals and underground water which constitute a unitary whole by virtue of their physical relationship.

2. For the purposes of this protocol "SADC", "the Council", "the Secretariat", "the Tribunal", and any other term defined in the Treaty establishing SADC shall have the same meaning as ascribed to tem in that Treaty.

Article 2

General Principles

For the purposes of this Protocol the following general principles shall apply:

1. The utilisation of shared watercourse systems within the SADC region shall be open to each riparian or basin State, in respect of the watercourse systems within its territory and without prejudice to its sovereign rights, in accordance with the principles contained in this Protocol. The utilisation of the resources of the watercourse systems shall include agricultural, domestic, industrial, and navigational ses.

2. Member States undertake to respect and apply the existing rules of general or customary international law relating to the utilisation and management of the resources of shared watercourse systems and, in particular, to respect and abide by the principles of community interests in the equitable utilisation of those systems and related resources.

3. Member States lying within the basin of a shared watercourse system shall maintain a proper balance between resource development for a higher standard of living for their peoples and conservation and enhancement of the environment to promote sustainable development.

4. Member States within a shared watercourse system undertake to pursue and establish close cooperation with regard to the study and execution of all projects likely to have an effect on the regime of the watercourse system.

5. Member States within a shared watercourse system shall exchange available information and data regarding the hydrological, hydrogeological, water quality, meteorological and ecological condition of such watercourse system.

6. Member States shall utilise a shared watercourse system in an equitable manner. In particular, a shared watercourse system shall be used and developed by member States with a view to attaining optimum utilisation thereof and obtaining benefits therefrom consistent with adequate protection of the watercourse system.

7. Utilisation of a shared watercourse system in an equitable manner within the meaning of paragraphs 4 and 6 requires taking into account all relevant factors and circumstances including:
   a. geographical, hydrographical, hydrological, climatical and other factors of a natural character;
   b. the social and economic needs of the member States concerned;
   c. The effects of the use of a shared watercourse system in one watercourse state on another watercourse state;
   d. Existing and potential uses of the shared watercourse system;
   e. Guidelines and agreed standards to be adopted.

8. Member States shall require any person intending to use the waters of a shared watercourse system within their respective territories for purposes other than domestic use or who intends to discharge all types of wastes into such waters to first obtain a permit from the

relevant authority within the State concerned. The permit shall be granted only after such State has determined that the intended discharge will not have a detrimental effect on the regime of the watercourse system.

9. Member States shall, without delay, notify other potentially affected States and competent international organisations, of any emergency originating within their respective territories.

10. In the event that implementation or execution of any planned measures is of the utmost urgency in order to save life, or to protect public health and safety, or other equally important interests as a result of an emergency situation, the Member State planning the measures may, notwithstanding the provisions of paragraph 9, immediately proceed with implementation or execution, provided that in such event a formal declaration of the urgency of the measures shall be communicated to the other Member States.

11. Member States shall take all measures necessary to prevent the introduction of alien aquatic species into a shared watercourse system, which may have detrimental effects on the ecosystem.

12. Member States shall maintain and protect shared watercourse systems and related installations, facilities and other works in order to prevent pollution or environmental degradation.

13. Shared watercourse systems and related installations, facilities and other works shall be used exclusively for peaceful purposes consonant with the principles enshrined in the SADC Treaty and in the Charter of the United Nations and shall be inviolable in time of international as well as internal conflicts.

Article 3

Establishment of River Basin Management Institutions for shared Watercourse Systems in the SADC region

1. Member States hereby undertake to establish appropriate institutions necessary for the effective implementation of the provisions of this protocol.

2. Without prejudice to paragraph 1 above, Member States undertake to establish the following institutions:
    a. A Monitoring Unit, based at the SADC Environment and Land Management Sector.(ELMS).

b) River Basin Commissions between Basin States and in respect of each drainage basin;

c) River Authorities or Boards in respect of each drainage basin.

Article 4

Objectives of the River Basin Management Studies

The River Basin Management Institutions shall have as their main objectives:

(a) To develop a monitoring policy for shared watercourse systems;
(b) To promote the equitable utilisation of shared watercourse systems;
(c) To formulate strategies for the development of shared water course systems;
(d) To monitor the execution of integrated water resource development plans in shared watercourse systems.

Article 5

Functions of the River Basin Management Institutions

In order to attain the objectives set out in Article 4, the River Basin Management Institutions shall, in consultation with watercourse States, perform the following functions:

(a) With regard to National Water Resources Policies and Legislation;

i. Harmonisation of national water resources policies and legislation,
ii. Monitoring compliance with water resource legislation and, where necessary, recommending amendments thereto and the introduction of new legislation.

(b) With regard to Research, Information and Data Handling;

i. Collecting, analysing, storing, retrieving, disseminating, exchanging and utilizing data relevant to the integrated development of the resources within shared watercourse systems and assisting member States in the collection and analysis of data in their respective States,
ii. Reviewing the provisions of National Development Plans relating to the water course systems,
iii. Designing and conducting studies, research and surveys relating to the environmentally sound development and management plans for shared water course systems,
iv. Stimulating public awareness and participation in the sound management and development of the environment including human resources development,
v. Promoting in accordance with the national development plans of the Basin States, the formulation of integrated master plans for shared watercourse systems.

(b) With regard to Water Control and Utilisation in shared watercourse systems,

i. Recommending regulation of the flow and drainage,
ii. Promoting measures aimed at flood and drought mitigation,
iii. Recommending and promoting measures to control desertification, soil Erosion and sedimentation
iv. Monitoring the utilisation of water for agriculture, domestic, industrial and navigational purposes,
v. Monitoring the establishment of hydro-electric power installations,
vi. Monitoring the generation of hydro-electric power,

(d) With regard to Environmental Protection;
i. Promoting measures for the protection of the environment and the prevention of all forms of environmental degradation arising from the utilisation of the resources of the shared watercourse systems,
ii. Assisting in the establishment of a list of substances whose introduction into the waters of a shared watercourse system is to be banned or controlled,
iii. Promoting environmental impact assessments of development projects within the shared water-course systems,
iv. Monitoring the effects on the environment and on water quality arising from navigational activities,

(e) With regard to Hydrometeorological Monitoring Programme;
i. Promoting a Hydrometeorological Monitoring Programme in consultation with other SADC sectors.

Article 6

Financial and Regulatory Framework for River Basin Management Institutions

A financial and regulatory framework for the River Basin Management Institutions referred to in Article 3 shall be annexed to this Protocol and shall constitute part of the Protocol.

Article 7

Settlement of Disputes

1. Any dispute between two or more member States arising from the interpretation or application of this Protocol, which cannot be settled amicably, shall be referred to the Tribunal for adjudication under Article 16 (1) of the Treaty of SADC.

2. (a) If a dispute pertaining to this Protocol is between SADC and a member State, a request shall be made by the Council for an advisory opinion in accordance with Article 16 (4) of the Treaty of SADC;

(b) The opinion given by the Tribunal shall be accepted by the parties as final and binding.

Article 8

Signature

This Protocol shall be signed by duly authorised representatives of Member States.

Article 9

Ratification

This Protocol shall be ratified by the signatory States in accordance with their constitutional procedures.

Article 10

Entry into Force

This Protocol shall enter into force thirty (30) days after the deposit of the instruments of ratification by two thirds of the Member States of SADC.

Article 11

Accession

This Protocol shall remain open for accession by any member State of SADC.

Article 12

Amendments

1. An amendment to this Protocol shall be adopted by a decision of three quarters of the Summit members who are party to this Protocol.
2. Proposals for amendments to this Protocol may be made to the Executive Secretary by any Member State for preliminary consideration by the Council of Ministers, provided however that the proposed amendment shall not be submitted to the Council of Ministers for preliminary consideration until all Member States have been duly notified of it and a period of three months has elapsed after such notification.

Article 13

Withdrawal

1. Any member State may withdraw from this Protocol upon the expiry of six months from the date of giving a written notice to that effect to the Executive Secretary.
2. Such a State shall cease to enjoy all rights and benefits under this Protocol upon the withdrawal coming into effect, but shall remain bound

to its obligations hereunder for a period of twelve months from the date of withdrawal.

Article 14

Termination

This Protocol may be terminated in accordance with the provisions of Article 35 of the Treaty of SADC.

Article 15

Saving Provision

Nothing contained in this Protocol shall derogate or be construed to derogate from existing agreements entered into between two or more member States or a member State and a State that is not a member State concerning the utilisation of shared watercourse systems, provided that member States shall endeavour to give effect to such agreements and any rights acquired or obligations assumed thereunder in conformity with the general principles prescribed in Article 2.

Article 16

Annexes

Any agreement that may be entered into between two or more member States or between a member State and a State which is not a member State, concerning the utilisation of one or more shared watercourse systems shall be in conformity with the provisions of Articles 2, 3, 4 and 5 of this Protocol.

Such Agreement may be adopted as an Annex to this Protocol by a decision of two-thirds of the Summit members who are party to the Protocol.

Article 17

Depositary

1. The original of this Protocol and all instruments of ratification and accession shall be deposited with the Executive Secretary of SADC, who shall transmit certified copies to all member States.

2. The Executive Secretary shall register this Protocol with the Secretariats of the United Nations Organisation and the Organisation of African Unity.

REFERENCES

1. Howe, C "Guidelines for the Design of Effective Water Management Institutions Utilizing Economic Instruments" Paper presented at UNEP Workshop on the Use of Economic Principles for the Integrated Management of Freshwater Resources, Nairobi June 1995.

2. Hufschmidt, Maynard M., James David E., Meister, Anton D., Bower Blair T., and Dixon John A 1983. Environment, Natural Systems and Development - An Economic Valuation Guide. The John Hopkins University Press, Baltimore and London

3. Southern African Development Community (SADC). 2003 [Revised] Protocol On Shared Watercourse Systems In The Southern African Development Community. Botswana.

4. Sivalingnam, G. 1995. Pricing Water for Sustainable Development. - Paper presented at UNEP Workshop on the Use of Economic Principles for the Integrated Management of Freshwater Resources, Nairobi June 1995.

5. Sivalingnam, G. 1995 Personal Commication.

6. Smith, D and Rast, 1998 W. *Environmentally Sustainable Management and Use of Internationally Shared Freshwater Resources* in Watershed Management – Practice, Policies and Co-ordination. Edited by Reimold, R. J. McGraw Hill New York.

7. United Nations 1993. Agenda 21 United Nations New York.

8. United Nations, 1997. Convention on the Law of the Non-navigational Uses of International Watercourses. United Nations, New York.

9. United Nations Environment Programme 1993. The Aral Sea - Diagnostic Study for the Development of an Action Plan for the Conservation of the Aral Sea. Technical Report, Freshwater Unit, Nairobi

10. United Nations Development Programme. 1995. The Aral In Crisis. Report presented at Tashkent, Uzbekistan.

11. United Nations Environment Programme 1999, Global Environmental Outlook 2000 UNEP/Earthscan, Nairobi/London.

12. United Nations Environment Programme 2002, Global Environmental Outlook 3 UNEP/Earthscan, Nairobi/London.

13. World Bank. 1993. Water Resources Management. Technical Report. Washington DC

ON THE PROBLEM OF THE CASPIAN SEA LEVEL FORECASTING

BOLGOV, M.V., FILIMONOVA, M.K. & M.D. TRUBETSKOVA
Water Problems Institute, Russian Academy of Sciences, Moscow, Russia

ABSTRACT

Stochastic fluctuations of climate and hydrological regime caused by both natural and anthropogenic factors are the main reason for the big uncertainty of long-term hydrological forecasts. Consequently, they cause the necessity to reconsider the risk of economic activities at inland sea coasts towards its increase. To estimate such a risk some sources of uncertainty arising under the sea hydrological regime forecasting are considered in the paper. By use of digital models of a region, some features of morphometric characteristics (depending on the sea level) are revealed, and their contribution to the level regime variability is appreciated.

INTRODUCTION

The forecast of long-term fluctuations of an inland water body level such as the Caspian Sea is the most complicated geophysical problem. It demands knowledge both of features of .the hydrometeorological regime of the region, and of mechanisms for the occurrence of long-term climate and runoff fluctuations [3]. Their definition is associated with different types of mistakes arising because of unreliable measurements, imperfection of the modeling representations, the approximate character of hypotheses, limitation of access to observation materials of recent years, etc.X

MODELX

Fluctuations of an inland (closed) water body level represent poorly predicted natural phenomenon, which nevertheless can be described on the basis of stochastic models of hydrometeorological processes and representations about its water balance.

Fluctuations of a water level (*h*) of a closed water body can be described with the help of the known water balance equation [4,5]:

$$\frac{dh}{dt} = \nu(t) / F(t) - e(t), \qquad (1)$$

Where $\nu(t)$ is water inflow per unit time, e(t) is an amount of effective evaporation (evaporation minus precipitation), F(t) is the water surface area.

To solve the equation (1), the linearized analogue is offered in [5]:

$$\frac{dh}{dt} + (b\,\nu(t)/a^2)h = \nu(t)/a - ae(t), \qquad (2)$$

H. Vogtmann and N. Dobretsov (eds.), Transboundary Water Resources: Strategies for Regional Security and Ecological Stability, 27–35.

Where a and b are the coefficients of linear dependence of water surface area on the level. Assuming the coefficient constant under h, we receive Langeven linear differential equation:

$$\frac{dh}{dt} + \alpha h = g(t), \qquad (3)$$

where $\alpha = b\bar{v} / a^2 = Const, \quad g(t) = v(t) / a - e(t)$.

Assume that at the initial moment $t=0$ the level equals to h_o relative to the so-called equilibrium value $\tilde{z} = \frac{1}{b}\left(\frac{\bar{v}}{\bar{h}} - a\right)$, where $\bar{v}$ and $\bar{h}$ are water inflow and evaporation expectations.

The solution of the equation (3) can be presented as [5]:

$$h_t = h_o e^{-\alpha t} + e^{-\alpha t} \int_0^t e^{-\alpha \xi} g(\xi) d\xi . \qquad (4)$$

Averaging the right and the left parts in (4), we receive

$$\bar{h}_t = h_o e^{-\alpha t}, \qquad (5)$$

as $\bar{g}(t) = o$.

The solution (4) of the equations (3) being considered as a linear operator transforming the random function g(t), the expression for the correlation function of the process h is received in [5] assuming the entrance process g $(R_{g(t)} = \sigma_g^2 e^{-\beta t})$ to be the Markov one. This expression being too bulky, we limit ourselves here only by the formula for the dispersion of level fluctuations:

$$\sigma_h^2(t) = \sigma_g^2 \{(\beta - \alpha) - (\alpha + \beta)e^{-2\alpha t} + 2\alpha e^{-(\alpha+\beta)t}\} / \alpha(\beta^2 - \alpha^2) \qquad (6)$$

At $t \to 0$ the following expression is true for the correlation function

$$R_h(t) = (\beta e^{-\alpha t} - \alpha e^{\beta r})/(\beta - \alpha). \qquad (7)$$

For the dispersion:

$$\sigma^2_h = \sigma^2_g / \alpha(\alpha+\beta). \qquad (8)$$

THE DATA

From the beginning of regular observations over the Caspian Sea regime, during about one century its levels were insignificantly fluctuating near the mark of -26 m. In 30th years of 20th century a catastrophic decrease of the level on 1.7m occurred. Further, the decrease of the level continued but much more slowly and in 1979, the level has reached -29m. The increase begun after that was observed until 1995, annual average levels exceeding -27 m.

One of the most important problems of the data analysis is the representativity of the existing observation series of the Caspian Sea level and the reliability of the statistical conclusions received on the limited data.

Researches of fluctuations of river runoff and evaporation from water body surface and the precipitation analysis have shown that the so-called simple Markov chain can be accepted as the mathematical model of these processes [1, 6, 8.]. Simulation of the Caspian Sea level series executed on the basis of the corresponding numerical algorithms [7], allows to make a conclusion, that probability distribution in an interval from 0,1 up to 99,9 % is well approximated by the normal distribution law. At the same time, it is necessary to note, that the histogram of the observed sea levels sequence has the two-modal form.

HYPOTHESIS OF THE STATIONARITY OF CLIMATIC CONDITIONS

Along with other hypotheses, one can find the explanation of the Caspian Sea abnormal behavior in a context of a climatic change problem. Climatic conditions are known to be essentially non-stationary on long time intervals (more than centuries). For example, according to some estimations, during last post-glacial period the World Ocean level has grown on 130 m. Instrumental measurements have demonstrated ocean level growth approximately on 15 - 20 cm for 100 years as well. However, this figure lies within the limits of measurement accuracy and can hardly serve as the evidence of essential modern climatic changes (or their indicator).

The indirect characteristic allowing estimating of the "stationary" hypothesis acceptability is the average duration of time when the sea level is above or below the set level (occurrence probability of series of years with extreme level values). It is found from the outlier probability distribution for the prescribed stochastic model. The solution of this problem for the Caspian Sea has shown [6] that the recurrence of long series (up to 50 years and over) in relation to the gravitation level is essential. The existing stochastic runoff fluctuation model seems to be advanced in the framework of some quasi-stationary theory by the account of long-term tendencies in the process of the Caspian Sea basin humidation, but such models have not been offered yet.

Recently, the trends discovered in wind speed on the coastal stations called the hypothesis of the climatic condition stationarity in question. The evaporation value is known to depend on wind speed as well as on air temperature and humidity. The lowered evaporation observed in last decades would be logically associated with fluctuations of these climatic characteristics. The researches carried out by a number of scientists have not revealed any of significant tendencies in air and water temperature and air absolute humidity. As for the module of a wind speed, the conclusion about the presence of a negative linear trend for the period 1960-1990 has been made [2].

On the basis of the observations on meteorological stations in the Caspian region, average monthly values of air temperature and surface wind speed for the period of 1961-1999 has been received for Izberg, Makhachkala and Tuleniy. These stations are located at the western coast of the Caspian Sea closely to each other. Thus, if any tendencies in change of climatic characteristics take place all over Caspian region, they should be shown equally at these stations.

On Fig.1 graphs of surface wind speed module for these three meteorological stations for January, April, July and October are presented.

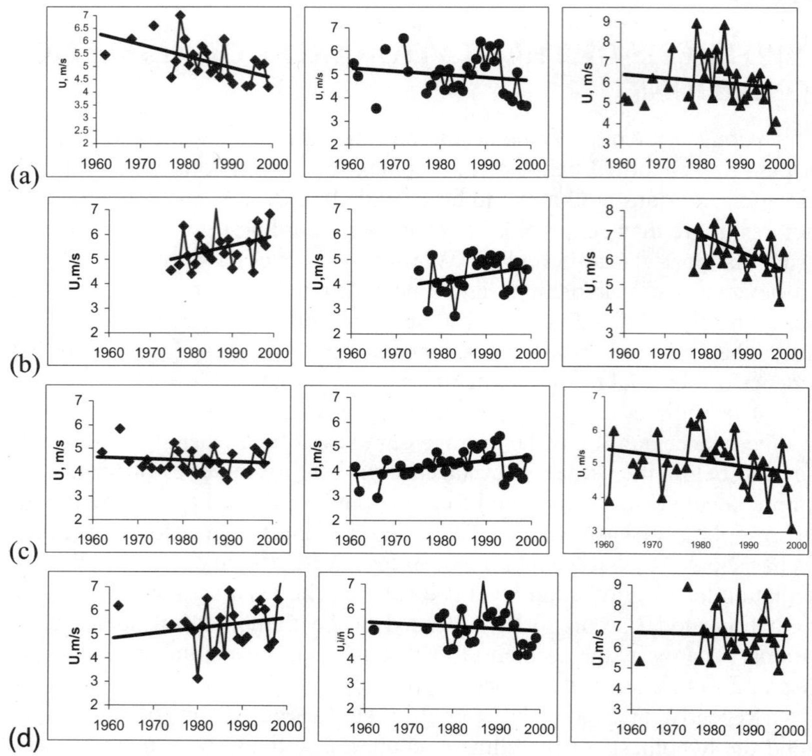

Fig.1. Long-term changes of wind speed module on stations: Tuleniy (♦), Izberg (•), Makhachkala (▲) for January (a), April (b), July(c), October(d), 1960-1999

At Tuleniy station, one can see the negative trend in surface wind speed in January; in April and to a lesser degree in October the positive one is observed, while in July no essential change is revealed. At Izberg station, on the contrary, in January and October no any essential change is observed, in April and in July an obviously expressed positive trend is observed. At Makhachkala station in January and October wind speed is practically the same for the given period, in April and - to a lesser degree - in July wind speed decreases. Thus, at three closely located stations wind speed turns to behave differently. It is necessary to notice as well that during the considered period differences in meteorological observations carrying out on the stations took place. So, at station Makhachkala during the period up to 1968 observations were carried out 4 times a day, then up to 1986 – 3 times a day (9, 15, 21h.), then again 4 times a day. Up to 1993, the observations were carried out at 3, 9, 15, 21h, from 1994 - at 0, 6, 12 and 18h. This could lead to fluctuations of average wind speed within the limits of 0.5 m/s that corresponds to approximately 10 % of average norm.

Thus, the analysis of change for last 40 years does not give us an opportunity to draw a conclusion about the existence of strongly pronounced tendencies of the meteorological characteristics that would determine evaporation from the Caspian Sea and changes of its level. On the contrary, the researches carried out can be regarded as the support of the hypothesis of climatic conditions stationarity.

MORPHOMETRIC DEPENDENCES

Some other ideas explaining the "anomaly" of the Caspian Sea are connected with mistakes of the accepted modelling representations. So, in linearized differential water balance equations linear dependence of the water surface area upon the sea level is used (the so-called morphometric dependence). Modern level of computer facilities allows us not to limit ourselves by opportunities of standard topographical maps when searching morphometric dependences. To solve such problems, it is necessary to automatize the access to elevation data. With this purpose, the relief digital model (a matrix of altitudes with a geographical fixation) was used with the grid step of 30 seconds, with smooth approximation along height. Calculations of the Caspian Sea morphometry are represented further without taking into account the Kara-Bogas-Gol gulf water area. The Caspian Sea is divided into three parts: Northern (to the north of 44°30' N), Middle (from 40°N up to 44°30' N) and Southern (from 40° N to the south).

Let us consider the distribution of areas occupied by various bathymetric steps. The most significant part of the area - 66.6 % - has the depths less than 100 meters, 42.4 % of it (28.2 % of the total sea area) located mainly (70 %) in the Northern part of the Caspian Sea having the depth less than 10 m. Depths more than 900 m occupy about 1 % of the area. The reminder area is distributed rather regularly between 200 and 800 m depths approximately by 4-5 % per 100m of depth. The general character of depths distribution is well seen on the bathygraphic curve of the sea (fig.2 A). One can see two smooth breaks at the depths of 500 and 100 meters and various inclinations: very flat one in the upper part, very steep one in the middle and less steep one in the lower part of the curve.

If to consider the part of the bathygraphic curve corresponding to the heights of -40 to -20m abs in more details (Fig. 2 B), a presence is obvious of a bathygraphic curve excess at about -28 m abs. It is the Northern part of the sea that is responsible for the excess, while curves of the Middle and the Southern parts of the Caspian Sea have no peculiarities within these altitudes. Such behavior of a curve is explained by high flatness of the relief in the coastal zone and of the Northern Caspian Sea coast. These relief features do not allow using linear interpolation in the field of a coastal zone area for the problems of forecasting of the sea level change and the coastal zone flooding.

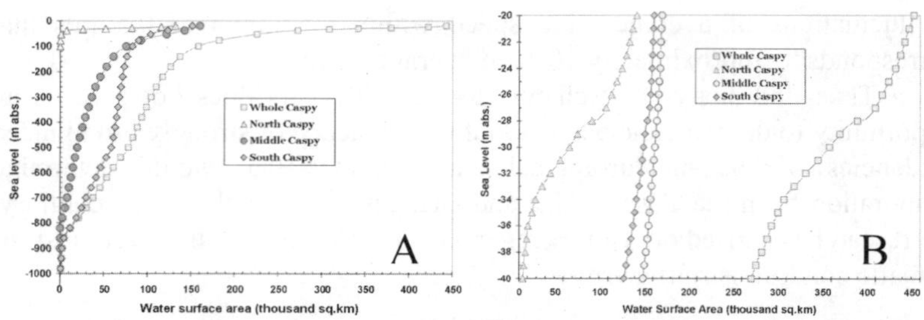

Fig. 2. Dependence of the water surface area (km^2) of The Caspian Sea on the level (m abs.). ***A*** *– for levels from –1000 to 0 m abs.,* ***B*** *- for levels from –40 to -20 m abs.*

The account of "new" morphometry of the Caspian Sea in problems of the long-term level forecast results in the underestimated values (quantile) with small excess probability compared with the "linear" problem.

THE FORECAST

With the modern level of scientific knowledge, it is impossible to make the forecast (long-term) of water balance hydrometeorological components for concrete calendar date. Hence, the method of the long-term calendar forecasting of a sea level is impossible as well. Only probability forecasts are possible, for example, the one of the average sea level position and the deviation from it of the position of the given probability (quantile).

In Table 1, the quantils of the conditional distributions of level probability for the nearest decades are presented. As follows from this table, the range of possible level values is wide enough. The mark - 26 m should be taken into account as one having the exceedence of 1 % when designing engineering protection actions. Low sea levels on marks - 28,-29 m are also probable.

Year Probability of exceedence	2001	2003	2005	2010	2020	2030	2040	2050
0,1%	-26,53	-26,11	-25,87	-25,59	-25,48	-25.48	-25,54	-25,63
1%	-26,66	-26,37	-26,20	-26,04	-26,02	-26,06	-26,13	-26,21
5%	-26,77	-26,59	-26,50	-26,44	-26,51	-26,58	-26,66	-26,73
Average	-27,05	-27,14	-27,22	-27,40	-27,67	-27,83	-27,92	-27,98
95%	-27,32	-27,69	-27,93	-28,37	-28,84	-29,07	-29,19	-29,23
99%	-27,43	-27,91	-28,23	-28,76	-29,33	-29,59	-29,71	-29,75

Table 1. Probability forecast of the Caspian Sea level (irrevocable withdrawals = 25 km^3/year; the initial level=-27.0m)

For the periods until 2005, 2010, 2020, 2030 etc. the average forecasting level (with 50 % probability of exceedence) is from -27.05 m (that practically corresponds to the modern coastal line position) up to -27.98 m for 2050 (actually up to the level, which is considered safe). The most adverse forecast with 0.1 % probability of exceedence for the same periods is -25.48 m, and the most adverse forecast with 1 % probability of exceedence is -26.02 m.

BAYES FORECAST ESTIMATIONS

The risk in economic development of coastal territories arises both as a consequence of stochastic character of influences as mid-annual or extreme levels and owing to a wide set of uncertainties, that should be taken into account when accepting some design (or organizational) decisions. The models presented above take into account the basic kind of uncertainty - probability character of inducing hydrometeorological processes. At that, parameter errors resulting from estimating by short samples are not considered.

Let us consider the influence of sample properties of model parameters estimations. Being functions of time, expressions for sea level expectation and dispersion enable to predict future fluctuations of the sea in the form of confidence intervals or the given probability values (the conditional density being approximated by the Gaussian law). The latest form of the forecast is used at a substantiation of actions and designing of constructions of coastal territories engineering protection. We remind that expressions (5) and (6) have been received in the assumption that estimations of stochastic models parameters are known exactly and the received conditional distributions reflect only stochastic character of hydrometeorological processes variability. However, those parameters estimations are known to possess the so-called sample properties and to be characterized by errors in the simplest case.

The simple approach leading to results easy to be interpreted is based on Bayes ideology supposing the construction of the so-called forecast density of required value x as the conditional distribution $\pi(x/y)$ with the given observations of y.

In accordance with the terminology [9], let us introduce the probability model for *x* as *g(x/θ)* dependent on some parameter θ determined by available values of y. Further, assuming that posterior distribution density of this parameter *p(θ/y)* is known and *x* and *y* are independent, the forecast probability density can be received from the following expression:

$$\pi(x \mid y) = \int_{\theta} g(x \mid \theta) p(\theta \mid y) d\theta \qquad (9)$$

Calculations according to the equation (9) are carried out by numerical integration with either sample distribution of a parameter (estimation), or the distribution of an estimation on homogeneous objects group (water bodies, lakes, meteorological stations, etc.) used as the distribution density *p*. As it was mentioned above, as *g*- function it is possible to use the normal distribution law with parameters determined by formulas (5) or (6) depending on parameters estimations of stochastic models of river inflow and water body evaporation. The results of the calculations are presented in Tables 2 and 3 for cases when the sample dispersion of estimations of inflow and evaporation average values and the autocorrelation coefficient estimation are accounted.

Year / Probability of exceedence	2001	2003	2005	2010	2020	2030	2040	2050
0,1%	-26.53	-26.11	-25.86	-25.56	-25.40	-25.36	-25.40	-25.48
1%	-26.66	-26.36	-26.20	-26.01	-25.97	-25.97	-26.02	-26.10
5%	-26.77	-26.59	-26.50	-26.42	-26.47	-26.51	-26.58	-26.65
Average	-27.05	-27.14	-27.22	-27.40	-27.68	-27.83	-27.92	-27.98
95%	-27.32	-27.69	-27.94	-28.39	-28.89	-29.14	-29.27	-29.31
99%	-27.43	-27.92	-28.24	-28.79	-29.39	-29.68	-29.82	-29.86

Table 2. Probability forecast of the Caspian Sea level with the account of sample properties (error) of the inflow expectation estimation (irrevocable withdrawals = 25 км^3/year; an initial level = -27.0m)

Year / Probability of exceedence	2001	2003	2005	2010	2020	2030	2040	2050
1%	-26.6	-26.36	-26.19	-26.00	-25.94	-25.94	-25.96	-26.01
99%	-27.43	-27.92	-28.24	-28.80	-29.40	-29.70	-29.86	-29.92

Table 3. Probability forecast of the Caspian Sea level with the account of the sample properties (errors) of the inflow and evaporation expectation estimations (irrevocable withdrawals = 25 км^3/year; an initial level = -27.0m)

CONCLUSIONS

The problem of the Caspian Sea level forecasting is closely connected both with the research of natural hydrometeorological processes variations and with the transboundary character of this water object. The changed status of the sea has led to essential degradation of the observation network and, correspondingly, to the growth of hydrological forecasts uncertainty and zones of risk.

As the result, the conclusion is obvious about the necessity of close international cooperation of scientists in the Caspian region with participation and under the support of UNESCO, UNEP, etc.

REFERENCES

1. Bolgov, M.V. (1994) Modelling of multivariate random variables by the method of expansion by natural orthogonal functions., Meteorology and hydrology, № 7, 82-85, (in Russian).
2. Dziuba, A.V. and Panin G.N. (2003) Contemporary changes of wind speed vector and the intensity of evaporation from the Caspian Sea Surface, Water Resources, V.30, 2, 198-207 (in Russian).
3. Kritsky, S.N. (1973) Methods of analysis and calculation of fluctuations of the closed water bodies level. Water resources, № 6, 9-26 (in Russian).
4. Kritsky, S.N. and Menkel, M.F. (1964) Fluctuation of the closed water bodies level. Works of Hydroproject, Moscow, № 12, 29-61, (in Russian).
5. Muzylev, S.V., Privalsky, V.E. and Ratkovich D.J. (1982), Stochastic models in engineering hydrology, Nauka, Moscow (in Russian).
6. Ratkovich D.Ya. (1976) Long-term river runoff variations. Gidrometeoizdat, Leningrad (in Russian).
7. Ratkovich D.Ya. and Bolgov M.V. (1994) The investigations of probability regularities of long-term fluctuations of The Caspian Sea level. Water resources, № 6, 389-404, (in Russian).
8. Sarmanov O.V. and Sarmanov I.O. (1983). Basic correlation types used in hydrology, Nauka, Moscow (in Russian).
9. Handbook on applied statistics. (1990) Eds. E.Llojd, U.Lederman, Moscow, Finance and statistics, (in Russian).

REGIONAL EXPERIENCES IN WATER RESOURCES PROBLEM SOLVING IN KYRGYZSTAN

MAMATKANOV, D.
Institute of Water Problems, Bishkek, Kyrgyzstan

ABSTRACT

This paper briefly analyses experience of the Central Asian countries in solving problems associated with transboundary water resources.

The paper shows that over the last decade the co-riparians have made several attempts to adjust their upstream-downstream relations. Different interstate agreements and water treaties could be considered as proof of these efforts. However, as the last several years demonstrated, none of these agreements or treaties has been observed in full. Opposite attitude towards issues related to water allocation, regional water strategy and sharing maintenance/ operational expenditures has prevented the upstream and downstream countries from solving water-related problems.

The paper puts forward some useful recommendations that might contribute to solution of the problems in focus.

The involvement of Kyrgyzstan in regional water management issues is explained by two important circumstances. In general, these circumstances are associated with trans-boundary waterways, which originate within the country's territory, and water-related facilities of interstate significance. Water resources of Kyrgyzstan are formed within boundaries of the following hydrologic basins (rivers: Syrdarya, Amudarya, Tarim, Chu and lakes: Balkhash and Issyk-Kul). All rivers of the country, except rivers of the Issyk-Kul lake basin, are of a trans-boundary nature. The average long-term drainage is about 50 km^3. Water allocations for Kyrgyzstan make up about 11 km^3 of the total drainage. In reality, the country utilizes 5-6 km^3, and the rest of water goes to neighboring states.

In the Central Asian region, the water allocation scheme of the Soviet period has still remained in force. This presents the main problem for the upper riparian states of the Aral Sea basin.

The essence of this problem lies in the fact that allocation of water resources among states of the Aral Sea basin had been carried out within a single state, the Soviet Union, and in accordance with principles of equal water supply for the existed irrigated areas. Preferences in land resources development were given to flat areas of the region (Kazakhstan, Uzbekistan). Within these zones foundations of cotton and rice industries of the Soviet Union had been laid. Water infrastructure has been built on trans-boundary rivers within territories of mountain countries (Kyrgyzstan, Tajikistan). Construction of this infrastructure was accompanied by

H. Vogtmann and N. Dobretsov (eds.), Transboundary Water Resources: Strategies for Regional Security and Ecological Stability, 37–42.

planned limitations restraining development of irrigated agriculture, mining and processing industries of these countries. Their economic stability was ensured by supplies of agricultural products and energy resources from the neighboring republics. Such approaches of the planned economy had negatively affected economies of the mountain republics. Negative impacts become particularly evident with the disintegration of the Soviet Union, suspension of subsidiary supplies and increase in export prices for energy resources to the market level by the downstream countries.

In 1992, a Treaty on cooperation in the field of joint management, use and protection of Water Resources was signed. The treaty confirmed and remained in force the former principles and rules, while the nature of interactions between the Central Asian countries changed from inter-republic to interstate. All states of the region considered the treaty as an interim measure valid until a new strategy of water allocation (with due attention to economic conditions of each republic) was developed. More than 10 years passed but only little has been done to solve this problem.

Why is the Kyrgyz Republic not satisfied with the existing scheme of water allocation? The country's water quota is 25% of total water resources formed within its territory.

Firstly, such quota is obviously insufficient to provide the existing irrigated areas with adequate water supplies. A great number of lands in the country are short of water.

Secondly, the existing water allocation system restricts opportunities for irrigated agriculture development as a source of secure food independence in arid zone conditions. The level of land resources use designated for agricultural purposes is evaluated by the UN Commission for Sustainable Development and is an indicator of arable land per capita. In Kyrgyzstan this indicator makes up 0.3 ha. As a result of such low index, the country is considered to be located in a zone of unsustainable land use. According to estimations, if the population growth rate in the country is 1.4 %, the level of arable lands per capita will reduce to 0.2 ha by 2025. This means that Kyrgyzstan will be in a zone of unsafe land use, and probability of food independence loss will increase. Taking into account the land degradation rate, the country may turn out in the zone of unsafe land use 5-7 years earlier. In spite of the existing significant land potential, expansion of new arable lands is limited due to mountain landscape. Under such circumstances, the Kyrgyz Republic needs to develop irrigated agriculture to provide its food independence. This development requires increase in water resources use within the country. Having 2 million ha of additional land fund, suitable for irrigation, the country's water consumption will total 20 km^3.

And thirdly, the existing water allocation scheme of the Soviet period puts limitations on the operation regime of the lower Naryn HPPs. This regime was reasonable in the context of the planned economy, when Kyrgyzstan was provided with inter-republican energy supplies to compensate electric power underproduction during the winter months. At present time, Kyrgyzstan fulfils its obligations under the 1992 Treaty and ensures the operating regime of the Toktogul reservoir adopted by the treaty. Uzbekistan and Kazakhstan in their turn provide Kyrgyzstan with energy resources on account of electric power produced during irrigational discharges. However, this barter exchange does not compensate all losses estimated at US $ 200 million.

These three circumstances confirm that without increase of water quotas, Kyrgyzstan will be unable to achieve economic stability and further sustainable development.

There is one more important problem, which is associated with new economic realities. Kyrgyzstan, at the expense of its budget, carries out exploitation of water regulating facilities. These facilities are important for the region since they provide neighbouring states with water supplies. Besides, the facilities have monitoring and water reproduction services. I think that in market economy conditions, a considerable part of maintenance and operational expenditures should be shared by all co-riparians. Fairness of such approach is quite clear for representatives of the Western countries presenting here. As for our region, like in case with the 1992 Treaty, the problem of shared maintenance and operational expenses has remained unsolved even after 11 years of market reforms.

Consequently, although several agreements and documents have been developed and signed, yet the co-riparians do not have any experience in solving the above-mentioned problems. Among such documents, there was the Aral Sea Action Plan adopted by heads of the Central Asian states and Russian Federation in 1994 in Nukus. It was generally aimed to develop a number of specific measures to improve ecological situation in the Aral Sea basin during the coming 5-7 years with due attention to the regional social and economic development. The primary task of the Action Plan was development of a new regional strategy of water allocation, rational water use and protection of this life-giving source. It was expected that this strategy would help to prepare drafts of interstate legal and standard acts regulating joint water use and the resource's protection. The Aral Sea Action Plan was followed by the Bishkek Declaration emphasizing the necessity to develop a new strategy of water allocation and economic mechanisms for water use.

The main provisions of the new regional water strategy (1997) have been developed within the framework of the Aral Sea Action Plan. These provisions, through approaches' determination, have brought us nearer to the strategy development itself. However, all works have been cut down. Recently, SPECA has started to work in this direction; yet its work is just in the initial stage that deals only with and discusses the developed strategy on rational and effective use of water and energy resources.

Today we can only talk about the existing insignificant experience in partial problem settlement. This includes the 1998 Treaty on joint and comprehensive use of water and energy resources of the Naryn – Syrdarya cascade signed by governments of Kazakhstan, Kyrgyzstan and Uzbekistan. Every year volumes of water supplies under this treaty are revised. I do not agree with such approach because our estimations show that it is unprofitable for Kyrgyzstan. The country does not receive more than US $ 100 million annually.

In my opinion, the 2000 Treaty on interstate use of water facilities of the Chui and Talas rivers signed by governments of the Kyrgyz Republic and Republic of Kazakhstan is more progressive in terms of its fairness. According to this treaty, Kazakhstan has agreed to share costs related to the operation of such water facilities.

What are initiations of Kyrgyzstan on solution of water-related problems?

As for legal aspects, the country has developed and adopted a number of laws regulating relations between the basin states with regard to trans-boundary water resources:

- Law “On Water” and
- Law “On interstate use of water objects, water resources and water facilities of the Kyrgyz Republic”.
- These two laws contain, among other things, the following principles and norms of the state policy with respect to water resources originated within the country’s territory:
- Recognition of the state’s ownership right to water objects, water resources and water facilities originated its territory;
- Recognition of water as a natural resource that has its definite economic value;
- Introduction of water pricing in interstate relations.
- In order to put into practice these principles and norms, an Interdepartmental Commission has been established under the government of the Kyrgyz Republic. In general terms, this Commission has been aimed at drafting a national policy on the use of trans-boundary water resources. The Commission continues its work, and I would like to use this opportunity to emphasize once again that our policy is based on the necessity to:
- Redistribute water resources with due attention to constitutional rights of newly independent states to their own natural resources;
- Share costs associated with monitoring, reproduction, and protection of water resources as well as with direct water supplies through water regulating facilities;
- Compensate regime losses of Kyrgyzstan, which are related with (i) the winter underproduction of electricity by HPPs due to irrigational discharges to satisfy needs of the downstream countries, (ii) loss of agricultural lands during construction of reservoirs of interstate significance.
- Only such approach can help to achieve the regional security and environmental stability.
- In case water quotas of the upstream countries are increased, the expected results will be as follows:
- The countries will have an opportunity to expand arable lands to provide their populations with secured food supplies;

- Reduction of water supplies to the lower riparian countries will promote rational water use and contribute to improvement of environmental situation in these areas. This is so, because more than 20 thousand m^3 per hectare currently consumed downstream lead to damage, rather than to benefit.

What share of water allocations does Kyrgyzstan claim to?

27.6 km^3 of water resources of the Naryn-Syrdarya basin, the most disputable basin in the region, is formed within the territory of Kyrgyzstan. The country's water quota is 4 km^3 (for irrigation of 450 thousand ha). In reality, the country's land resources available for irrigation make up 1.3 million ha. Taking into account one of the principles of the Helsinki Convention (adequate provision of next generations with water resources), we assume that the Kyrgyzstan's water quota should be estimated as a sum of volumes for irrigation of the above-mentioned area and perspective industrial/ municipal/ household water use. In other words, the country needs approximately 14 km^3 or 50% of the total water resources formed within its territory.

As for cost sharing, we think that it would be much convenient to share expenditures related to interstate water flow delivery on basis of the water pricing approach. This method proposes two types of water tariffs: (i) tariff for water as a natural resource and (ii) tariff for regulation and supply services.

The first type of tariffs includes expenditures for monitoring, flood protection/ coast strengthening/ reforestation measures in the headwaters zones.

The next type of tariffs contains expenditures associated with operation and maintenance of reservoirs and hydro-technical constructions of interstate significance. Advantage of these tariffs is obvious for all co-riparians since the generated funds could be used to improve monitoring systems, water level forecasts, water resources reproduction, flood/ landslide protection measures and maintenance of dam/ hydro-technical constructions. All these arrangements could provide security in the downstream zones. It should be also mentioned that access to monitoring information will be open. Besides, preconditions necessary for the establishment of a joint information center will be created.

All existing disagreements on the working regime of the Toktogul reservoir can be solved by means of (i) compensation for the winter underproduction of electricity by HPPs due to irrigational discharges to satisfy needs of the downstream countries and (ii) compensation for loss of agricultural lands during construction of reservoirs of interstate significance.

The total entitlement payment makes up about US$ 200 million. During the summer months the Uzbekistan's profit of using waters of the Naryn river totals US$ 800 million. In case the upper riparian country duly receives compensation of all losses, profitability of the proposed approach will be clear: the downstream countries will be provided with water by a specified date and in the fixed quantity.

In conclusion, I would like to thank to the symposium organizers for the given opportunity to participate in this event. Let me also express my hope that our countries will do their best to solve the discussed problems.

REFERENCES

1. Asanbekov A., Sahvva K. and Shapar A. The Economic Mechanism for Transboundary Water Resources Management and the Main Provisions of the Interstate Water Allocation Strategy. Bishkek: International Institute of Mountains, 2000.

2. Bazhanova L., Romanovski V. and Dikikh A. "Water And Hydropower Resources of Kyrgyzstan." In Mountains of Kyrgyzstan. Bishkek: Technologiya Publishing House, 2000.

3. Erdman O. "Concept of Water Resources Complex Use and Protection." Bulletin of the International University of Kyrgyzstan, No 4/12 (2000): 38-85.

4. Shapar A., Erdman O. and Asanbekov A. "Methodology for Estimation of Annual Losses Caused to Kyrgyzstan by Construction of the Toktogul Reservoir and Its Operation in Irrigation Mode." Proceedings of the Academy of Sciences of the Kyrgyz Republic, No 4 (1998): 51-55.

REGIONAL EXPERIENCE IN SOLVING PROBLEMS OF WATER RESOURCES IN UZBEKISTAN

RAKHIMOV, SH. KH.
Institute "SANIIRI", Tashkent, Uzbekistan

Keywords: river basin, water resources, surface water, groundwater, return water, water control, water reservoirs, solution of water problems, improvement of water management

ABSTRACT

During the last ten years, water resources problems became particularly urgent in the Central Asian region. To solve that problem, heads of the five Central Asian states' water-economical ministries had meetings with each other and signed some agreements on coordinated water resources management. Therefore, in each republic, specifically in Uzbekistan, works conducted towards the improvement of water resources control.

The Aral Sea basin, along with ones of Nile, Ind, Khuankhe and Mesopotamia, is one of the most ancient civilization centres of world history which have been risen based upon irrigated agriculture and water resources development.

Two main rivers of the basin, Amudarya and Syrdarya, have been giving off their waters to the irrigation of arid desert oasis and to the Aral Sea from ancient times on. Data of average long-term natural flow of the rivers is stated in the report.

At the present, five independent states are situated on the basin: the Republic of Kazakhstan, the Kyrgyz Republic, the Republic of Tajikistan, Turkmenistan and the Republic of Uzbekistan, with a total population of about 40 million people.

Concerning the Aral Sea basin:

Renewable water resources – 120 km^3 of water a year;

Area of the basin – 1550 thousand km^2;

Irrigation area– 7.95 M ha;

Gross output – \$ 79 B a year;

Electricity production – 126 B kWh, including hydroelectric power stations – 34.5 B kWh;

Total number of reservoirs having capacity more than 100 M m^3 – 80;

Total capacity of the reservoirs – 60 km^3;

Total length of the irrigation network:

H. Vogtmann and N. Dobretsov (eds.), Transboundary Water Resources: Strategies for Regional Security and Ecological Stability, 43–54.

Inter-farm one – 47.75 thousand km;
In-farm one – 268.6 thousand km;
Total number of vertical drainage wells – 865;
Total length of collector-drainage network – 191.9 thousand km,
Including underdrainage -collector– 47.9 thousand km;
Multipurpose water-resources scheme includes:
Unique water reservoirs with capacity 19 km^3, such as Toctogul;
Dams of the Charvak, Andijan and Nurek hydro systems with height of 100-350 m;
The biggest all over the world gravity canal Karakumskiy with discharge of 600 m^3/s and length of 1400 km.
A unique cascade of machine canals including the Karshy cascade with discharge of 350 m^3/s and lifting height of 180 m.

After a series of low-water years in the 1980s, some serious difficulties related to water resources management in the Aral Sea basin arose amongst the Central Asia republics and Kazakhstan. To settle and eliminate the collisions, the former Ministry of water resources of the USSR established two basin water-management associations (BWMA) named “Amudarya” and “Syrdarya”.

The Aral Sea’s available water resources are formed by renewable surface water and groundwater of natural origin, as well as return water of anthropogenic origin.

SURFACE WATERS

All the water resources pertain, to the most part, to the Syrdarya and Amudarya river basins. Many centuries ago, independent basins (inland ones, but trending to the Amudarya river) formed by the Kashkadarya, Zarafshan, Murgab, Tedjen, lost their link to the main river.

Basin of the river	River runoff being formed within the state					Total of the Amu-darya basin
	Kyrgyz Republic	Tajiki-stan	Uzbe-kistan	Turk-meni-stan	Afgha-nistan and Iran	
1	2	3	4	5	6	7
Pyanje	-	21.089	-	-	13.200	34.289
Vakhsh	1.604	18.400	-	-	-	20.004
Kafirnigan	-	5.452	-	-	-	5.452
Surkhandarya	-	0.320	3.004	-	-	3.324
Kashkadarya	-	-	1.232	-	-	1.232
Zarafshan	-	4.637	0.500	-	-	5.137

Murgab		-	-	-	0.868	0.868	1.736
Tedjen		-	-	-	0.560	0.561	1.121
Atrek		-	-	-	0.121	0.121	0.242
Rivers of Afghanistan		-	-	-	-	6.743	6.743
Total of the Amudarya basin	km^3 %	1.604 2.0	49.9 62.9	4.736 6.0	1.549 1.9	21.593 27.2	79.28 100

Table 1. Natural runoff in the Amudarya river basin (average longtime runoff within 1934 – 2002 years, km³ a year)

Basin of the river	River runoff being formed within the state				Total of the Amudarya basin
	Kyrgyz Republic	Kazakhstan	Tajikistan	Uzbeki-stan	
1	2	3	4	5	7

Table 2. Natural river runoff in the Syrdarya river basin (average longtime runoff within 1951 – 2002 years, km³ a year)

Naryn	14.544	-	-	-	14.544
Karadarya	3.921	-	-	-	3.921
Rivers of the					
country between				0.312	2.072
Naryn and					
Karadarya ones	1.760	-	-	0.408	1.188
Right bank of					
the Ferghana					
valley					
Leftt bank of	0.780	-	-	0.190	4,545
the Ferghana					
valley					
<u>Midstream</u>					
<u>rivers Chirchik</u>	3.500	-	0.855	0.145	0.295
Akhangaran					
Keles	-	-	0.150	4.100	7.949
Arys and Bugun	3.100	0.749	-	0.659	0.659
Downstream	-	-	-	-	0.247
rivers	-	0.247	-	-	1.183
	-	1.183	-	-	0.600
	-	0.600	-		
Total of the Syr Darya basin, km^3	27.605	2.426	1.005	6.167	37.203
%	74.2	6.5	2.7	16.6	100

Amudarya is the biggest river of Central Asia. Its length from Pyanje's origins is 2540 km, and the basin area is 309 thousand km^3. After the junction of Pyanje with Vakhsh, the river is called Amudarya. At the half of the stream, there are three big inflows from the right (Kafirnigan, Surkhandarya….) and one from the left (Kunduz), joining the Amudarya. Further up to the Aral, there isn't any inflow. River feed is mostly formed by melted snow and glacial waters, therefore maximum discharges are to be observed in summer, and minimum – in January and February months. Such an annual distribution of flow is extremely favourable for the use of river water for irrigation. Running through the plain from Kerki towards Nukus cities, Amudarya looses the most part of its flow to evaporation, infiltration and irrigation. The turbidity of the river is the highest in Central Asia, and also one of the highest in the world. The major flow of Amudarya is formed on Tajikistan Territory. Further, the river runs along the frontier between Afghanistan and Uzbekistan, then crosses Turkmenistan and returns to Uzbekistan (just "returns to Uzbekistan") where it (finally) enters the Aral Sea.

The Syrdarya river is the longest river of Central Asia and it has the second largest water content in this region. Its length from the Naryn's sources is 3019 km, and the basin area is 219 thousand km^2. Syrdarya's sources are in Central (Inner) Tyan-Shagn. After the junction of Naryn to Karadarya, the river is called Syrdarya. Spring – summer tide is typical for this water regime, which begins in April. The greatest flow occurs in June. The major flow of Syrdarya is formed on Kyrgyz Republic territory. Then Syrdarya crosses Uzbekistan and Tajikistan and enters the Aral Sea on Kazakhstan territory.

Dividing flows of Amudarya and Syrdarya according to zones of forming within the country has been made by means of GIS. The data cited (table 3) shows that for the Kyrgyz Republic 25.1 % of the total flow of the Aral Sea basin is formed, in Tajikistan – 43.4 %, in Uzbekistan – 9.6 %, in Kazakhstan – 2.1 %, in Turkmenistan – 1.2 %, and in Afghanistan and Iran – 18.6 %.

State	River basin		Aral sea basin	
	Syrdarya	Amudarya	km^3	%
1	2	3	4	5
Kazakhstan	2.426	-	2.426	2.1
Kyrgyz Republic	27.605	1.604	29.209	25.1
Tajikistan	1.005	49.578	50.583	43.4
Turkmenistan	-	1.549	1.549	1.2
Uzbekistan	6.167	5.056	11.223	9.6
Afghanistan and Iran	-	21.593	21.593	18.6
Total basin of the Aral Sea	37.203	79.280	116.48	100

Table 3. Total river natural flow in the Aral Sea basin (average long-term flow, km^3 a year, Research-engineer center of ICWC estimate)

UNDERGROUND WATER

Renewable resources of groundwater in the Aral Sea basin can be divided by their origin into two sorts: it can be formed naturally in mountains and on a catchment territory and or it can be formed under the influence of filtration on irrigated areas. On the whole, 339 water deposits have been explored and approved to use on the basin territory. In total,

43.49 km^3 of groundwater reserves have been evaluated in the region; thereof 25.09 km^3 in the Amurdarya basin and 18.4 km^3 in the Syrdarya basin. Mostly, groundwater deposits have sufficiently strong hydraulic interdependence with the surface flow. This becomes apparent by the reduction of surface flow because of the excessive withdrawal of groundwater. Taking into account this, and also according to the capacity of wells equipped by each deposit, state committees have approved reserves that are allowed to withdraw. Total quantity of allowed reserves are 16.94 km^3 (see table 4). Total current groundwater withdrawal for various water users is 11.04 km^3 a year, although in the beginning of 1990s it exceeded 14.0 km^3.

State	Estimate of regional reserves	Allowed reserves for use	Actual with-drawal in 2002	Used to					
				Drinking water supply	Industry	Irrigation	Vertical drainage	Experi-mental exhausts	Others
Uzbekistan	18455	7796	7749	3369	715	2156	1349	120	40
Total of the Aral Sea	43486	16938	11037	4307	1088	4045	1409	121	67

Table 4. Groundwater reserves and their use by states within the Aral Sea basin (million km^3 a year)

Major part of the basins' groundwater is situated and formed on the territory of two countries and is thus transboundary water (Golodnaya Steppe's, Kyzylskiy, Dalverzinskiy, Kafirniganskiy, Ferganskiy and so on). As water withdrawal rises and water shortage increases in these regions, the regulation, control and international licensing to avoid its exhausting will become more and more a problem as well as deleterious effects, contamination and to ensure steady water use in the future. Unfortunately, these issues have been disregarded by water management and local administration bodies of the region's countries so far.

RETURN WATERS

Return waters are an additional source of water available to use in the Aral Sea basin. But, considering their high mineralization, these are at the same time the main polluters of water bodies and the environment. Around 95 % out of the total amount of return waters are collector-drainage waters from irrigation; the rest is sewage from industrial and utilities.

As irrigation and the building of drainage systems develop in the region, we can find a continual rise of return water, which was especially intense within the 1960-1990 years' period. In 1990s, return water account stabilized and even began to decrease somewhat because of stopping irrigation development, degradation of drainage systems, and starting realization of measures on water saving. On average for the 1990-2000 years' period the total return water was varying between 28.0 km^3 and 33.5 km^3 a year. About 13.5 –15.5 km^3 has been formed annually in the Syrdarya basin and about 16 - 19 km^3 in the Amudarya basin (see the table 5). More than 51 % of the total return water is exported through collectors to rivers, around 33 % - to lowlands. Only 16 % of return water is used repeatedly for irrigation because of the uselessness of this water as a result of pollution.

State	Collector-drainage water from irrigation*	Sewage from industrial and public service	Total of return water being formed	Water export and utilization		
				Into rivers	To natural lowlands	Repeated use to irrigation
Uzbekistan (total), including:	18.4	1.69	20.09	8.92	7.07	4.1
Basin of Surdarya	7.6	0.89	8.49	5.55	0.84	2.1
Basin of Amudarya	10.8	0.8	11.6	3.37	6.23	2
Total in the Aral Sea basin, including:	29.55	2.9	32:45	16.77	10.87	4.81
Basin of Surdarya	11.95	1.44	13.39	9.16	1.54	2.69
Basin of Amudarya	17.60	1.46	19.06	7.61	9.33	2.12

* Allowing for exhausting by vertical drainage holes

Table 5. Forming return water and water export in the Aral Sea basin (average for a period of 1990-1999), km^3 a year

State of use and management of return water represents a big problem until now and in the future, which lays beyond the sphere of regional, and more often, national organizations.

Great water amount discharged into rivers without any limitation turns good freshwater into poor-mineralized water, which is hardly used for any need. Ponds in desert zones and at outskirts of irrigated lands are irregularly fed with collector-drainage water. As a result, these ponds lose their ecological and natural stabilizing significance. In the region, few hundreds reservoirs of various capacity and size have been built based on collector-drainage and wastewaters. Notable among them are the Aydar-Arnasay lowland with a capacity of more than 20 km^3, Sarikamysh with a capacity of about 100 km^3, Denghizkul, Solenoe, Sudochye and several small ones containing a few millions of cubic meters. In the reservoirs, as a rule, there isn't any flower or fish production. Fauna and Flora are not able to develop, because of the unstable water- salt- regime that is formed without any control, under the influence of fortuity factors

FLOW REGULATION BY WATER RESERVOIRS.

In the Aral Sea basin, more than 60 reservoirs have been constructed and are in use. Total full capacity of the reservoirs amounts to 64.5 km^3, therefrom 46.5 km^3 is the conservation zone, including 20.2 km^3 in the Amudarya basin and 26.3 km^3 in the Syrdarya one (see the table 6).

Name	Date of putting into operation	Full capacity	Dead zone	Source
		Million cubic meter	Million cubic meter	
The Amudarya river basin				
Uzbekistan				
Tujamujun	1980	7800	2550	Amudarya
Tudakul	1986	1200	50	Amu-Bukhara Main
Talimarjanskoe	1978	1525	125	Canal
Yujno-	1962	800	210	Karshy Main Canal
Surkhanskoe	1985	500	30	Surkhandarya
Tupolangskoe	1978	170	17	Tupolang
Shurkulskoe	1957	320	80	Zarafshan
Kuya-Mazarskoe	1989	130	20	Amu-Bukhara
Akdarjinskoe	1941	840	24	MainCanal
Kattakurganskoe	1984	53	3	Akdarya
Karaultjubinskoe	1957	29.5	5.7	Zarafshan
Kamashinskoe	1961	55	15	Zarafshan
Kattasajskoe	1967	260	17	Yakkabagdarya
Pachkamarskoe	1983	27.2	3	Katta-Say
Dekhcanabadskoe	1959	425	0	Guzdarya
Chimkurganskoe	1982	170	15	Kichik-Uradarya
Ghissarakskoe	1959	160	80	Kashkadarya
Uchkizilskoe				Aksu
				Zang canal

	Total	14464.7	3244.7	
The Syrdarya river basin				
Uzbekistan				
Djizakskoe	1968	100	4	Sanzar
Zaaminskoe	1979	51	21	Zaaminsu
Charvakskoe	1966	2000	420	Chirchik
Tuyabuguzskoe	1959	250	26	Akhangaran
Akhangaranskoe	1971	260	30	Akhangaran
Farkhadskoe	1947	350	330	Syrdarya
Kasansayskoe	1942	165	10	Kasansay
Karkidonskoe	1963	218.4	4.4	Kuvasay and South Fergana Canal
Kurgantepinskoe	978	33.3	5.5	Shakhimardan
Andijanskoe	1978	1900	150	Karadarya
	Total	5327.7	1000.9	

Table 6. Reservoirs in Uzbekistan

Due to the reservoirs which have been built, the regulation level (firm water yield) of the flow in Syrdarya is 0.94 (i.e. natural flow is nearly fully regulated), and in Amudarya – 0.78 (i.e. there are reserves for future regulation).

As a result of a series of low-water years in the 80s, some serious difficulties related to the water resources management in the Aral Sea basin arose amongst the Central Asia republics and Kazakhstan. To settle and eliminate the conflicts, the former Ministry of water resources of the USSR established two basin water-management associations (BWMA), named "Amudarya" and "Syrdarya".

After the collapse of the USSR, the heads of the water resources ministries from the five Central Asian states met on February 18th, 1992 in the city of Alma-Ata to avoid the re-arising of the former conflicts. As a consequence of discussions, meetings and negotiations, they signed the "Agreement on unified water resources management" and an agreement on the establishment of a united body – the Interstate Coordination Water Committee (ICWC), including the two BWMAs, a Secretariat and Research Information Centre (RIC), which are provided with executive functions of the ICWC. Thus it is possible to realize a united water policy which takes into account the population's interests and fields of the respective economies of the region's states; it also keeps the annual water supply of the states within limits and it provides water management, water use improvement, and the amelioration of the ecological situation in the region.

Nowadays, regional organizations of water resources management in the Aral Sea basin are the Interstate Fund of Saving the Aral Sea (IFSAS), including an Executive Committee and its affiliations in each state, in form of a founder/benefactor. The Executive Committee consists of the ICWC and the Interstate committee of social-economical, scientific-research and ecological cooperation with its institutes: the Secretariat and RIC. Existing organizational structure of the IFSAS is shown on the figure 1.

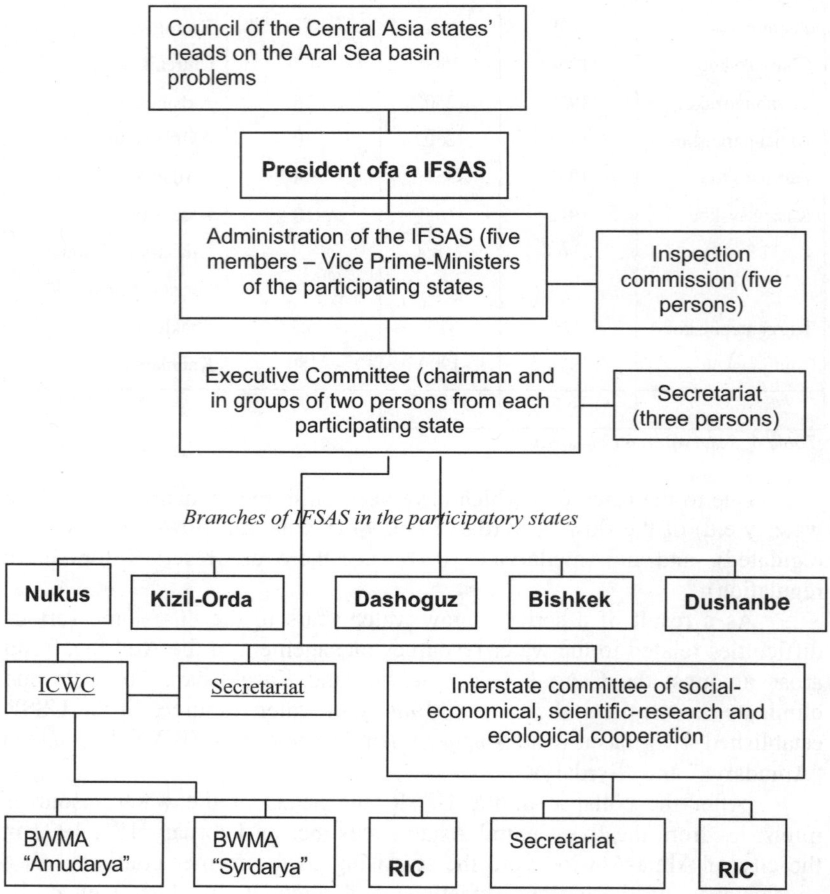

Figure 1. Existing organizational structure of the IFSAS.

In order to perfect the water resources management in the region and to improve the basin environmental ecology, and by grants of the World Bank and the European Union, the following agreements between the region's countries have been worked out under the direction of the leading research institute of the ICWC – SANIIRI, and the participation of research and design institutes of all states of the region:

1. Agreement on organizational structure of unified management, conservation and water resources development in the Aral Sea basin.
2. Agreement on use of water resources at the present conditions.
3. Agreement on joint planning use, development and conservation of transboundary water resources.
4. Agreement on transboundary water conservation, regulations of controlling their quality, and ensuring ecological stability in the region and others.

These agreements aim at an improvement of the organizational structure of the regional bodies controlling water resources; regulating use of transboundary waters of all kinds (surface, groundwater and return water) by the region states; planning use, development and conservation of the transboundary waters; securing ecological stability in the region. The region states' governments now are considering these documents.

In Uzbekistan, some sections of the Law "On water and water use" have been revised and upgraded in order to ensure that they fit to the regional requirements of water resources functioning.

In the republic, improvement measures for the arrangement of water economy management started according to the transition of the administrative-territorial method for the water resources control towards a basin management for irrigation systems.

11 basin administrations have been established which are charged with:

- Assessment, goal and efficient use of surface and return water resources based on the introduction of market principles and mechanisms of water consumption;
- Carrying out of the united technical policy in water economy based on the inculcation of advanced water saving technologies;
- Providing of irrigation systems and water facilities with operational reliability.
- Reconstruction and modernization of hydro-technical constructions and their equipment, energy and communication systems for safe maintenance of irrigation systems;
- Implementation of up-to-date systems of monitoring water resource use;
- Implementation of informational water resources control systems on the basin territory to rise its efficiency;
- Dialogue with all basin water users relating to the decision-making process on water resources use;
- Efficient use of the basin water resources considering the needs of sectors of the national economy, environmental ecology and firm water supply to population.

At the same time, the restructuring of the republic's agriculture, from collective farms to cooperative farms has taken place. On the basis of the farms, Water User Associations (WUA) have been established, in order to hand over the water control functions on the level of inter-economical and farm-irrigation systems. Their number in the republic today is more than 70,000. In light of the establishment of the WUA, workings on attraction of water users' means to maintain irrigation systems and to set rates for water supply and water charge.

Agreements on the improvement of the organizational structure of the regional bodies of water resource control, on the use of transboundary waters and the assurance of ecological stability in the region have been developed. The connection of the national water law and regional requirements of water management functioning, the reorganization of the national water resources control bodies, the restructuring of agriculture and the establishment of WUAs are the obvious case of using the regional experience of water resources management when solving water resource problems in Uzbekistan.

REGIONAL EXPERIENCES IN SOLVING OF WATER RESOURCES PROBLEMS IN TAJIKISTAN

NORMATOV, I.
Prof., Institute of Water Problems, Dushanbe, Tajikistan

Keywords: Irrigation, Syrdarya, hydropower, vegetation, reservoir, Central Asia, Amydarya , Nurek

ABSTRACT

The principal spheres of water resources use in Central Asia today are irrigated agriculture and hydroelectric engineering.

Problem of mutual relations between irrigation and hydroelectric engineering in the region is determined by the fact that the countries of upper stream-Kyrgyzstan and Tajikistan are interested in the energy regime of river flowing use, and the countries of the lower stream-Kazakhstan, Turkmenistan and Uzbekistan are interested in the irrigative regime.

The only possible settlement of the water division problem in the region remains reconsideration of existed limits. As the world practice shows in today's conditions its limits and needs for water which is the most moveable, changeable element of interrelations between the countries.

The notion Central Asia (the former name is Middle Asia and Kazakhstan) that is used nowadays includes the republics of CIS: Kazakhstan, Kyrgyzstan, Tajikistan, Turkmenistan and Afghanistan. Hydrographically the region of Central Asia (CA) is distinguished as the Aral Sea basin, which in its turn consists of two basins – the Syrdarya and the Amudarya Rivers.

The main indicators of technical and economic development of Central Asian economic region are given in Table 1.

During the times of the USSR all Central Asian republics were considered as industrially developed countries. Today they are countries with transitional economy that greatly lowered their economic level. Alongside, a great difference in today's economic development of separate Central Asian countries can be noted. (Table 1).

H. Vogtmann and N. Dobretsov (eds.), Transboundary Water Resources: Strategies for Regional Security and Ecological Stability, 55–72.

Country	Territory, thousand km^2	Population, million people	Per capita gross inland output by purchasing capacity parity, thousand dollars/ man	Per capita energy consumption, tons of conventional fuel/man
Kazakhstan	2636.20	14.95	3.56	3.67
Kyrgyzstan	198.50	4.90	0.68	0.66
Tajikistan	143.10	6.20	0.99	0.84
Turkmenistan	488.00	4.70	1.52	3.30
Uzbekistan	447.36	24.60	2.26	2.70
CA	3913.16	55.35	2.22	2.64

Table.1. Indicators of macroeconomic development of Central Asian region

During the times of the USSR all Central Asian republics were considered as industrially developed countries. Today they are countries with transitional economy that greatly lowered their economic level. Alongside, a great difference in today's economic development of separate Central Asian countries can be noted. (Table 1).

One of the main resources of Central Asian region is water. Not without reason the year 2003 is announced as "International Year of Fresh Water" at the UNO General Assembly by the initiative of Tajikistan – the country, where more than 50% of all water resources of the region are formed. In August 2003 the International Water Forum will be held in Dushanbe, the capital of Tajikistan.

Total water resources of the Aral Sea basin surface waters are 115.6km^3/year (Table 2). According to approximate evaluation underground water resources in the Aral Sea basin are 43.7km^3/year, 15.8km^3/year (36.2%) of them being the approved exploitation stores. Moreover, a large quantity of return waters are formed in the Aral Sea basin – 45.8km^3/year, only a small part of which is used anew for irrigation – 6.0km^3/year, and a great part of water is led to rivers (23.5km^3/year) and natural reduction (16.3km^3/year).

Evaluation of these resources sufficiency for Central Asia is ambiguous. If we compare them with leading countries with similar climatic conditions, first of all with Israel, we can come to a conclusion that water resources available today are quite enough when using modern technology of water usage (Table 3).

On the other hand under conditions of management, existing in the region at present, an undisputable fact today is deficiency of water resources, especially in arid years.

Country	The Amudarya River basin, km^3/year	The Syrdarya River basin, km^3/year	The Aral Sea basin	
			km^3/year	%
Kazakhstan	—	4.50	4.50	3.9
Kyrgyzstan	1.90	27.4	29.30	25.3
Tajikistan	62.9	1.1	64.00	55.4
Turkmenistan	2.78	—	2.78	2.4
Uzbekistan	4.70	4.14	8.84	7.6
Afghanistan	6.18	—	6.18	5.4
CA	78.46	37.14	115.6	100.0

Table.2.Surface water resources of the Aral Sea basin

Indicators	Central Asia	Israel
Total specific consumption per capita, m^3/year	345.0	2875.0
Including on irrigation, m^3/ha	5590.0	12887.0
The same, taking into account natural precipitations, km^3/ha	10.390.0	14690.0

Table.3. Specific consumption of water in Central Asia and Israel

The most well known consequence of such situation is the Aral Sea crisis and degradation of the Aral Shore.

The principal spheres of water resources use in Central Asia today are irrigated agriculture and hydropower engineering.

Irrigated agriculture appearance in Central Asia refers to the 6th –7th century B.C. Since then up to nowadays its role has been constantly growing, irrigated areas have been increased. By the beginning of the 20th century about 3.5 mln ha have been already irrigated in the region. Intensive development of irrigation in the region especially began during existence of the USSR (mainly from the 60's up to the 90's of the past century).

As a result by 1990 total area of irrigated lands in the region has increased up to 8.8 mln ha, including:

- in Kazakhstan - up to 2.8 mln. ha
- in Kyrgyzstan - up to 1.1 mln. ha

- in Tajikistan - up to 0.7 mln. ha
- in Uzbekistan - up to 4.2 mln. ha

Similar sharp growth in the soviet period was observed in power engineering too. We can say that beginning from the 30's of the past century the whole people generation could observe the foundation of modern basic branch of economy – electro power engineering, that was quite new for it. Total established capacity of all electric power stations in the region by the middle of the 90's grew up to 37.8 mln. kW, including:

- in Kazakhstan - 18.5 mln. kW
- in Kyrgyzstan - 3.8 mln. kW
- in Tajikistan - 4.4 mln. kW
- in Uzbekistan - 11.3 mln. kW

At that time the capacity of hydropower stations in the region reached 11.31 mln. kW, including:

- in Kazakhstan - 2.22 mln. kW
- in Kyrgyzstan - 2.95 mln. kW
- in Tajikistan - 4.40 mln. kW
- in Uzbekistan - 1.74 mln. kW

Unfortunately all these impressive results led to the same great negative consequences. Intensity of processes of ecological equilibrium violation in the region, which especially became apparent in the Aral Sea zone, has sharply increased; lands salting and their desertification has grown; the quality of water especially in the lower stream of rivers has worsened. At that already by the 1970's water resources of the Syrdarya River basin were almost completely exhausted. All this practically turned into a global ecological problem of the region, and regarding the Aral Sea – into ecological disaster.

One of the reasons of this situation was that the programs of development of both irrigation and power engineering laid down at the times of the USSR were not accomplished.

First of all according to ecological, and in the first place according to economic reasons, the project of Siberian rivers transference into the region, which could cardinally solve all the problems of irrigation development and saving the Aral Sea even under preservation of existing conditions of management, was fully stopped.

Practically the program of hydroelectric engineering development in the region has just been started. Construction of new hydroelectric power stations with total capacity of 7,7 mln. KW and putting them into operation was provided for Tajikistan by "Conception of the USSR power engineering development for the period 1991 – 2005".

The problems became even more acute after the collapse of the USSR and formation of new independent states when they obtained intergovernmental status.

One of such problems is connected with contradiction between irrigation and hydroelectric engineering.

Irrigated agriculture demands maximum use of water during vegetative period from April to October. And hydroelectric engineering is concerned with paramount use of river flowing in winter, the coldest period of a year when rivers contain little water, from October to April. Thus by the irrigative regime filling reservoirs is necessary in winter and their use – in summer, and by the energy regime it is vice versa, filling reservoirs is necessary in summer and their use – in winter. It is impossible to combine their interests within one reservoir.

But such a situation just takes place in the Aral Sea basin today. The number of reservoirs is quite limited in the zone of flowing formation, where the principal regulation of flowing has to be carried out. There is only one such reservoir in Tajikistan in the Amudarya River upper reaches – the Nurek reservoir. In the Syrdarya River basin there are three such reservoirs: the Tocktogul reservoir in Kyrgyzstan, the Kayrakkum reservoir in Tajikistan and the Andijan reservoir in Uzbekistan. But from the latter three the Tocktogul reservoir can only carry out long-term regulation of flowing. Besides, all of them are located in different states, and co-ordination of their work is in itself a problem.

The common problem of mutual relations between irrigation and hydroelectric engineering in the region is determined by the fact that the countries of the upper stream – Kyrgyzstan and Tajikistan, are interested in the energy regime of river flowing use, and the countries of the lower stream – Kazakhstan, Turkmenistan and Uzbekistan, are interested in the irrigative regime. An extremely complicated situation in this regard arose in the Syrdarya River basin, where all water resources were exhausted already by the 1980's.

The peculiarity of today's situation is that it did not appear primordially at the formation of water - energy complex of the Aral Sea basin, but already in the process of its functioning as a consequence of cardinal changes of geopolitical and economic conditions in the region.

At the formation of water - economic complex of Central Asia during the times of the USSR all the questions regarding complex use of water resources were unambiguously settled within a common systemic approach, though to a considerable extent an administrative-command one:

- A diagram was established that performed financing hydro units of a complex function at the expense of different branches shares in proportions determined by economic calculation.
- Common and individual criteria of effectiveness were worked out and priorities were identified. In Middle Asia such a priority was primordially given to cotton- growing.
- And finally, necessary mechanisms of compensation that functioned at that time were worked out. In particular electricity losses of separate union republics connected with the work of hydro units in the irrigative regime were provided by intergovernmental off-seasonal electrical energy transmissions between HPS and TPS under their work regime organized correspondingly.

Nowadays after the collapse of the USSR and formation of new independent states such a diagram does not function. In all the states their own national interests advanced to the forefront.

At that neither the international law nor the national one and existing precedents provide for any terms or commitments regarding exploitation of their reservoirs by states-owners even when they are located on trans-bordering rivers. At best their code of conduct towards neighboring countries can be formulated in the following way:

A sovereign state possesses all the rights on absolute establishment of corresponding to its national interests any river flowing regulation regimes on the reservoirs belonging to it and located on its territory.

In case that these regimes damage interests of other states of the basin, a state-owner must on concordance change its work regimes in favor of other states concerned with provision it from its side with a proper compensation.

Taking into consideration complexity and length of such interrelations formation during the transitive period of newly independent states creation, special agreements providing for preservation of mutual relations existing between them during the times of the USSR in the field of water-energy resources were signed between Central Asian republics:

Agreement between the Republic of Kazakhstan, the Republic of Kyrgyzstan, the Republic of Uzbekistan and Turkmenistan on cooperation in the field of joint management of intergovernmental sources water resources use and protection. Alma-Ata, February 18, 1992.

Nukuss declaration of Central Asian states and international organizations on problems of sustainable development of the Aral Sea basin. Nukuss, September 20, 1995.

In the first of them it is written down: “Recognizing commonality and unanimity of the region water resources the Sides possess equal rights on use and responsibility for provision of their efficient use and protection”.

In the second it is declared: “We agree with the fact that Central Asian states recognize earlier signed and acting agreements, treaties and other acts corresponding to norm regulating mutual relations between them on water resources in the Aral Sea basin and take them for their unswerving execution”.

Signing these agreements was caused in that complicated period by new states formation, striving for not breaking, but smoothly reforming the system, for permitting of no anarchy, but providing succession in decisions, that was, undoubtedly, justified. At the same time they bore a political character without touching upon economic substance of the question. Therefore such a scheme of interrelations could not remain long without changes.

Apropos, not only possibility but also necessity of such changes was established in the already signed at the same time:

Agreement on joint actions on the Aral Sea and the Aral Shore problems, on ecological enhancement and provision of social-economic development of the Aral Sea. Kzyl-Orda, March 26, 1993.

In it it is noted: The states-participants recognize as common objectives: regulating the system and improving the discipline of water use in the basin, working out corresponding intergovernmental legal and normative acts providing for use of common for the region principles of recovering losses and damages".

This present-day scheme of interrelations between Central Asian republics began to be created already in 1994. In March 17,1998 it was registered officially by signing Bishkek "Agreement on use of the Syrdarya River basin water and energy resources between the Government of the Republic of Kazakhstan, the Government of the Republic of Kyrgyzstan and the Government of the Republic of Uzbekistan", which was joined by the Republic of Tajikistan in June 17, 1999.

A common scheme of interrelations between the sides regarding services and compensations at the Syrdarya River flowing regulation was determined according to this agreement. It provided that:

Extra electric energy generated in excess of needs of the Republic of Kyrgyzstan and the Republic of Tajikistan by the cascade of Naryn-Syrdarya hydro-electric power stations connected with the regime of removals of water into vegetation and a long-term flowing regulation in the Tocktogul and the Kayrakkum reservoirs is given in equal parts to the Republic of Kazakhstan and the Republic of Uzbekistan. Its compensation is according to concordance realized by supplies of energy resources in the equivalent volume (coal, gas, furnace black oil, electrical energy) as well as other production (work, services) or as pecuniary compensation to the Republic of Kyrgyzstan and the Republic of Tajikistan to make necessary annual and long-term reserves of water in reservoirs for irrigative needs. While performing reciprocal settlement of accounts a common tariff policy on all types of energy resources and their transportation must be provided.

Bishkek Agreement signed in March 17,1998 can be considered as absolute success. Unfortunately its practical implementation leaves much to be desired.

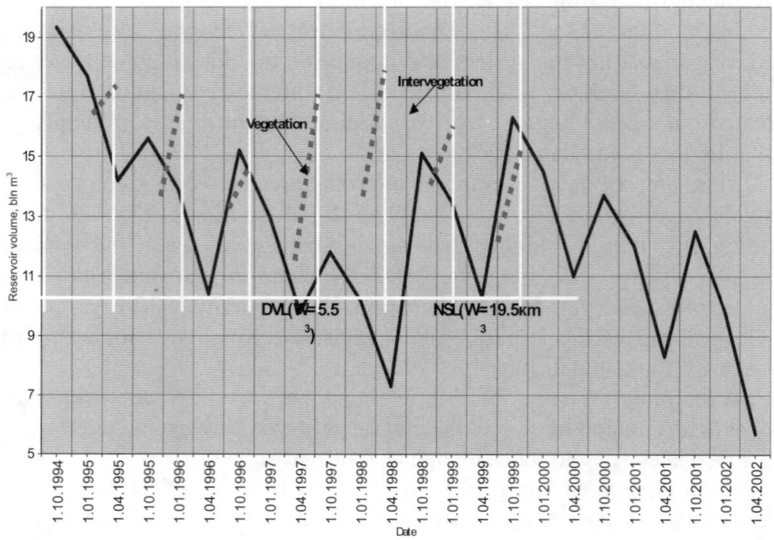

Figure 1: Diagram of the Tocktogul Reservoir

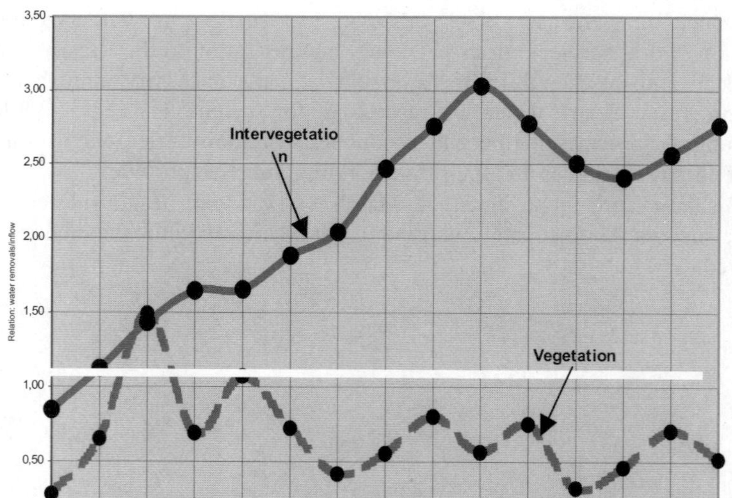

Figure 2: Regime of the Tocktogul Reservoir work

According to fig.1 and 2 the Tocktogul reservoir does not carry out seasonal regulation of flowing in irrigative purposes. It can be well seen that during the last years without exception the Tocktogul reservoir is filled in the vegetative period and used in the winter period, i.e. it works in the pure energy regime. The Tocktogul reservoir does not carry out a long-term regulation of flowing either. As a consequence from 1994 to 2002 the reservoir was nearly completely exhausted, although according to Table 4 deepness of the river during all this period was above the norm.

Inflowing km^3/year	Year								
	1 994	1 995	1 996	1 997	1 998	1 999	2 000	2 001	A ver
Annual	1 5,24	1 0,9	1 3,7	1 0,8	1 4,5	1 4,5	1 2,6	1 1,4	1 2,96
longterm Average	11,4								

Table.4. Inflowing to the Tocktogul Reservoir

Increased winter removals from the Tocktogul reservoir is in addition the reason of creating an unfavorable situation in the middle stream of the Syrdarya River that is lower than the hydro unit of Chardarya. As a result of the river's insufficient carrying capacity, the water flows into Arnasay cavity, by which the problem of the Aral Sea aggravates. And although the reason of such a situation is erection by Kazakhstan of partitioned off constructions in the Syrdarya riverbed, which is a violation of the acting norms and regulations of CNAR, this only corroborates the thesis that nowadays- ecological problems in the water energy complex are to a considerable extent connected with administrative organizational decisions.

Until recent times Tajikistan improved the situation to some extent by changing the regulation of river flowing by the Kayrakkum reservoir in the interests of the irrigative complex of Kazakhstan and Uzbekistan. This is well shown in diagrams on fig. 3 and 4.

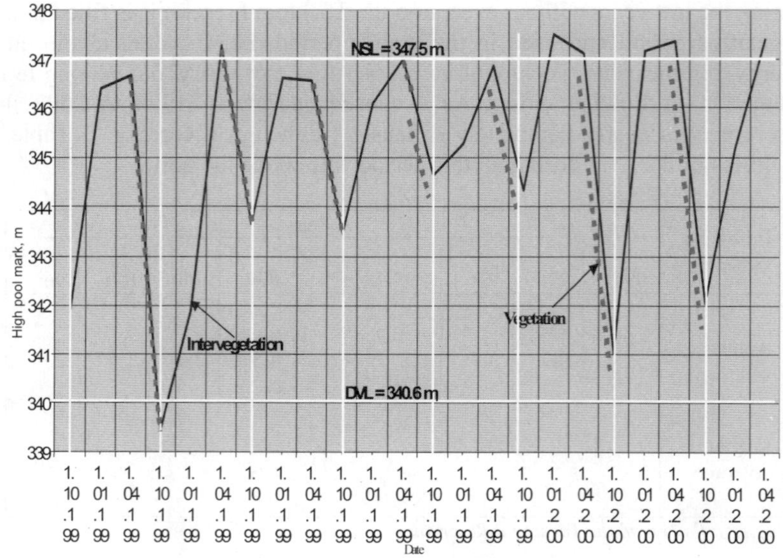

Figure 3: Kayrakkumreservoir. Diagram of seasonal filling and work.

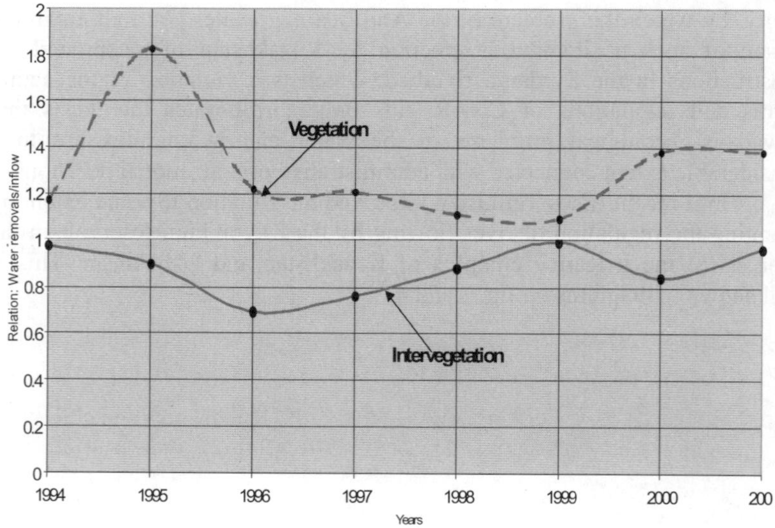

Figure 4: Regime of Kayrakkumreservoir work.

But first, in contrast to the Tocktogul reservoir the Kayrakkum reservoir can only carry out a seasonal but not a long-term regulation of flowing that is insufficient in a long-term section. Second, as it will be shown below, without getting sufficient compensation for its delivered services Tajikistan makes changes of a flowing regulation just by virtue of the developed tradition, displaying its good will. Such a scheme cannot work long in existing today market conditions.

But stable, sustainable use of water-energy resources of the basin is not provided even in this case. Working in the energy regime Kyrgyzstan annually uses volumes of the Tocktogul reservoir flowing that is greater in size than inflowing to it. As a consequence already today Kyrgyzstan made an official statement pointing out that Kazakhstan and Uzbekistan irrigative complex's needs in water can only be provided with the deficiency of 1,5 bln. m^3 a shallow year. At best this will lead to putting 150 thousand of irrigated lands out of use. And this is already a serious problem for the region. In actual fact the situation is more disquieting. According to the broadcast made by the Kyrgyz side at preserving this work regime the Tocktogul reservoir can be completely drained off to a dead volume already in two shallow years. When shallowness comes during this period the deficiency of water in the vegetative period will make 3-5 bln. m^3. This is already an ecological disaster. At the same time Kyrgyzstan itself incurs severe losses while using long-term water reserves in the Tocktogul reservoir. The point is not even that at such an approach at the coming of shallow years there will be no water left in the country itself. We can say that the regime of the reservoir use is not in itself optimal from the energy point of view, as the hydroelectric power station works at that with fewer pressures. The losses at that are very considerable. This is well shown in Table 5.

Reservoir vol., km^3	19,5	17,0	15,0	13,0	11,0	9,0	7,0	5,0
HPS pressure, M	174	164	157	149	140	130	120	108
1	2	3	4	5	6	7	8	9
Use of water on 1KW/h, m^3	2,29	2,42	2,57	2,68	2,87	3,20	3,46	4,16
Under-generation, mln KW/h	—	268	522	710	975	1377	1628	2164

Table.5. Technical-economic indicators of Tocktogul HPS according to the reservoir volume

Due to the work at the half-full reservoir with the volume 11-12 km^3 direct losses of Tocktogul HPS electricity generation in the years 1994-2001 make on average 950 mln KW/h a year, i.e. 20% of even the highest possible HPS generation.

However in spite of complete underfulfilment by Kyrgyzstan of its functions on a long-term and partly on a seasonal flowing regulation, Uzbekistan and Kazakhstan give full payment for its services according to the Agreement. Besides, Uzbekistan pays also Tajikistan for the services on seasonal flowing regulation by the Kayrakkum reservoir, i.e. it pays twice for one and the same service.

Payment to Kyrgyzstan for its services is carried out according to the following scheme. At the same time with additional water removals during a vegetative period in the volume of 2,2 bln m^3 Kyrgyzstan also annually supplies Uzbekistan and Kazakhstan with electrical energy surplus in total volume of 2,2 bln KW/h. The energy is distributed in equal parts between the countries-recipients.

In the frame of compensation, Uzbekistan

- Supplies Kyrgyzstan with 600mln KW/h of natural gas that is equivalent to 3,22 bln KW/h of electric energy at the work of HPS with thermo- and electric energy at 60% of coefficient of efficiency
- Supplies Kyrgyzstan with 20t of turbine oil and 500t of transformer oil. At their price making correspondingly \$400/t and \$300/t their total cost is equal to \$230 thousand that is equivalent to 11,5 mln KW/h of electric energy at its price 2 cents per KW/h.

- Delivers services on railway transportations for the sum equal to $500 thousand that is equivalent to 25 mln KW/h of electric energy at its price 2 cents per KW/h. Thus in total Uzbekistan returns Kyrgyzstan in the equivalent 3,25 bln KW/h of electric energy
- Kazakhstan supplies Kyrgyzstan by the same way of compensation:
- 566,7 t of Karaganda coal that is equivalent, at its use in TPS for generation of thermo- and electric energy at 60% of coefficient of efficiency, to 2,97 bln KW/h of electric energy even at the low calorific value of coal equal to 4500 kcal/kg.

Consequently as a result of all this, Kyrgyzstan has for itself a direct economic profit in the volume of 4,02 bln KW/h of extra electric energy.

The volume of the extra-received electric energy is greatly considerable for Kyrgyzstan. It makes one third of the total electrical energy production in the republic and only a little less than its annual generation in the biggest in the republic, the Tocktogul HPS.

At the same time together with water removals in a vegetative period Tajikistan annually supplies Uzbekistan with 300 mln KW/h of electrical energy.

By way of compensation the republic only received 200mln KW/h of electrical energy from Uzbekistan. Moreover, according to the agreement during three months when Tajikistan receives electrical energy from Uzbekistan, it does not present payment to the latter for the services on the regulation of frequency. In usual conditions such a payment makes about $250 thousand per month that makes $750 per three months. At the rate of 2 cents per KW/h it is equivalent to 337 mln KW/h of electrical energy. At that Tajikistan gets no compensation from Kazakhstan at all.

So receiving 200 mln KW/h of electrical energy from Uzbekistan Tajikistan gives it 337 mln KW/h, i.e. unlike Kyrgyzstan Tajikistan simply works at a loss.

In this way, as a consequence of all these imperfections all the participants of the Bishkek agreement signed in March 17, 1998 suffer appreciable losses:

- Uzbekistan and Kazakhstan do not have assured provision of water for irrigated agriculture especially in shallow years on account of absence of the long-term flowing regulation of the Syrdarya River by the Tocktogul reservoir.
- At the same time Uzbekistan, in the point of fact, pays for one and the same services on the seasonal flowing regulation twice – both to Kyrgyzstan and Tajikistan.
- At that underpaying a little Tajikistan these republics very considerably (more than twice) overpay Kyrgyzstan.
- Not carrying out a long-term flowing regulation Kyrgyzstan has, nevertheless, constantly to work the Tocktogul reservoir on account of winter deficiency of electrical energy, as a result of which the Tocktogul HPS always works on reduced

heads of water. Annual loss of electrical energy generation is equal to about 1bln KW/h. At that, unlike the compensation for the flowing regulation representing exchange of energy resources, this means direct losses.

- And finally, Tajikistan receives electrical energy from Uzbekistan by way of compensation for carrying out the seasonal regulation of flowing 1,5 times less than the quantity that it supplies itself. Kazakhstan does not pay Tajikistan for these services at all.

The main reason of this perplexed situation is absence of a clearly developed and coordinated economic mechanism of calculations of services and compensations for the regulation of flowing. Solution of this problem is a top-priority question for the republics of the Aral Sea basin today.

In principle different schemes are possible here. Apparently one of the best variants for this is the version of joint ownership. It could not only smoothly solve itself the question of the complex use of water flowing, but also serve the purposes of unification, integration of the states. Absence of such problems in the former USSR is explained by existence of common ownership then. It is just this joint ownership, but not ideology, politics and others that united the peoples of the USSR then, and it is just division of this property and, as a consequence, breakup of economic ties that is the reason of today's economic difficulties of the CIS countries.

Common property of Central Asian countries can be formed by building new projects and also as a result of admitting already acting ones into a joint-stock company.

Kyrgyzstan suggests that they should settle the question of river flowing joint use on the basis of commodity-market relations, which mean giving the water a commodity status and selling it to other states. Most likely this question does not have practicable prospects. First of all, fluvial running water is not an article of trade in the ordinary sense. It cannot be packed, marked or certificated. Its supply cannot be stopped on pure physical reasons. And finally, it is not absolutely clear how the transit of water as an article should be considered, co-coordinated and paid through the third countries, for example, by its delivery from Kyrgyzstan to Uzbekistan through Tajikistan, or to Kazakhstan trough Kyrgyzstan and Uzbekistan.

Suggested often palliative to water selling in the form of share holding in exploitation of water-economic projects is not sufficiently well grounded. Firstly because the result of functioning of any project is not only expenses on its exploitation, but also a profit from selling production and services made by it. Therefore it is incorrect to take into consideration only expenses without taking into account a profit.

Nowadays the most practicable and well-grounded version of the concordance of irrigation interests of the countries at rivers of the lower reaches and energy interests of the countries at rivers of the upper reaches is a scheme of compensations as it was provided in the Agreement on use of water-energy resources of the Syrdarya River signed in 1998, but in the elaborated form.

In a general sight this version looks like in the following way. The countries with zones of flowing formation as the base for calculation of the compensations work out a national regime of their hydro units work (Kyrgyzstan – for the Tocktogul, Tajikistan – for the Kayrakkum) without taking into account the interests of the countries located lower. Then they work out the second version of the same hydro units work, but this time taking into account the interests of the countries located lower. Economic difference between these two versions, losses and damages, connected with transition from the first version to the second one in a pecuniary or physical expression, determine the necessary volume of the compensations.

It is necessary in these compensations to take into account all losses, damages and expenses. This corresponds to a well-known principle of WTO "A user pays". This principle proceeds from the fact that in the price of natural resources all kinds of expenses should be taken into account that are connected with their use including expenses on liquidation of impact on environment in connection with exploitation, processing and use of the given type of resources.

Panel stated that the principles of "General agreements concerning the correction to the border tax" give a GATT country-member "an opportunity to follow the principle "a pollutant pays", but they are not obliged to this. Such a position of GATT can be explained by the fact that the principle "a pollutant pays" is discriminative for a national production and makes only it incur all expenses.

It is necessary to make another note. As a matter of fact a simple difference between these two regimes only determines prime cost of services on the regulation of flowing. In order to determine the price of these services it is necessary to add some norm of profitability to prime cost.

In conclusion it can be noted that there are also more cardinal settlements of this problem – concordance of the irrigation and energy interests. No matter how paradoxical it is it lies in greater development of hydroelectric engineering. The point is that today's contradictions between them are connected with the fact that there is only one great reservoir in every republic of upper stream: the Tocktogul in Kyrgyzstan and the Kayrakkum in Tajikistan. It is naturally that they cannot work in energy and irrigative regimes at the same time. And if there are more reservoirs then it will be quite possible to distribute their functions.

Another very significant problem of water-energy resources use in Central Asia is the problem of water division. It is the very problem that causes the sharpest arguments, and frequently, mutual suspicions between the republics. This problem is rooted in the Soviet past. Especially the most important thing is the necessity of its open discussion and working out common principles of settlement.

Today the positions of Uzbekistan, Kazakhstan and Turkmenistan in this question consist in the request to preserve the existing limits of water division and allocation of additional limits for the Aral Sea and the Aral Shore. The positions of Kyrgyzstan and Tajikistan consist in reconsideration of these limits with increase of their shares (not for today, in the perspective). At the same time Kyrgyzstan and Tajikistan demonstratively ground their requests on the increase of water resources limits by the fact that they were deprived by water division and did not get any compensation for this during the times of the USSR. As a consequence they possess now the least specific area of irrigated land per man in accordance with other republics, and they cannot even provide their population with the minimal level of consumption owing to their own agricultural production. Fairness of the requirements of the countries of lower stream on the necessity to increase water resources limits for the Aral Sea raises no doubts. Apropos Kyrgyzstan and Tajikistan were always concordant with them in this regard, as today's situation in the Aral Sea zone negatively impacts on them too. It is connected with dusty and salty winds from the territory of the former sea, which are spread up to glaciers and cause their intensive thawing. Separation of the Aral Shore as additional water user together with the Aral Sea itself causes objection. It is apparently simply an effort to increase its own limits. In order to exclude this and, besides, taking into consideration the fact that today there is not any reliable and objective control of water use inside separate republics, it may have sense not only to exclude the Aral Shore, but also the Aral Sea itself from the number of water users, and instead if this to set limits to Uzbekistan and Kazakhstan. And one, certainly, cannot agree with the equal responsibility of all the states for the Aral Sea destruction and with their equal participation in allocation of water limits. Such limits must be created first of all at the expense of the republics, which sharply reduced flowing into the Aral Sea in the 60's – 90's due to sharp increase of irrigated lands on their territory, i.e. owing to Kazakhstan, Turkmenistan and Uzbekistan. Both Kyrgyzstan and Tajikistan bear minimal relation to this.

Sometimes as an argument for preservation of the existing water division the republics of the Aral Sea lower stream raise the "historic" right. In our case such an approach cannot be recognized as grounded. In order to be really "historic" this right must, at least, be based on a long period as for example in case with Turkey, Syria and Iraq where the latter were guided in their demands by the four thousand existence of irrigation on their territories, beginning with the Shumers civilization, with approximately one and the same irrigation volumes. Unlike this, Kazakhstan, Turkmenistan and Uzbekistan are only guided in their requests by the situation of the 80's of the past century. In this way they make an attempt to secure as the "historic" right the maximum quotes that were achieved only once in the period, which was unambiguously recognized erroneous by all Central-Asian republics both now and in the times of the former USSR as regards water-energy resources use that led to ecological disaster of the Aral Sea.

Separate experts and first of all foreign ones suggest that they should settle contradictions that exist in the region in direct connection with the questions of water division limits by increasing efficiency of irrigative water use, increasing coefficient of efficiency of irrigation. Hopes on this are to a considerable extend exaggerated. Such water-economic sector reforming needs a lot of finances, and today both Kyrgyzstan and Tajikistan suffer shortage of them even for normal exploitation of water-economic objects. The example of Israel given mostly for grounding such an approach simply confuses. Israel can afford itself implementation of the most up-to-date technology as the country is on the highest level of economic development. Our republics cannot be compared with Israel, which has great opportunities regarding attracting foreign investments.

Thus the only possible settlement of the water division problem in the region remains reconsideration of the existed limits. And there is nothing unusual in this. As the world practice shows in today's conditions its limits and needs for water that are the most moveable, changeable element of interrelations between the countries. They are identified by certain conditions and depend on the reforms, development strategies, population dynamics and many others carried out in the states. Kazakhstan may serve as a good evidence for this. From 1998 to 2002 its need for water in vegetative period has already decreased from 1100mln m^3 up to 700mln m^3 in the Syrdarya River basin at the expense of carried out market reforms and connected with this reconsideration of structure of agriculture.

The necessity of changing the existing limits of water division is also connected with the situation in neighboring Afghanistan. After stabilization of the situation in the country it designated its request on water in the Amudarya River basin in the volume up to 25 km^3 a year. Today these volumes are not taken into consideration at all in water-economic balances of Central-Asian region.

And finally reconsideration of water division limits between Central-Asian states is simply necessary only because they are not legally assigned till present at all. Today's acting limits are established by protocols of scientific-technical councils of the USSR Ministry of Melioration and Water Economy in the 80's of the past year. Even at that time they bore not governmental but only departmental character. Nowadays when there is neither the USSR Ministry of Melioration and Water Economy nor the USSR, they have no juridical force. Of course we should realize that the reconsideration of limits is a complicated question that needs a very carefully approach. But on the other hand any attempt to cover it up will aggravate the situation more, and can finally lead to conflicts between the republics. And moreover requests on reconsideration of water resources limits from the part of Tajikistan, Kyrgyzstan and especially Afghanistan refer, as it has already been noted, not to the present time but to a distant perspective. Therefore there is still time today for analysis of the situation, consultations and talks, for preparation and conclusion of appropriate agreements between them.

We can also offer a relatively smooth mechanism of reconsideration of the existing water division in the profit of earlier deprived countries of Tajikistan and Kyrgyzstan. Additional limits of water for them can be received not at the expense of direct reducing the limits for other countries (Kazakhstan, Turkmenistan and Uzbekistan), but owing to economy of water by implementation of more efficient technology in these countries, and at the same time preservation of former irrigated areas.

The above-performed analysis of Central-Asian water-energy complex problems shows that on the whole the region possesses necessary resources and potential for normal sustainable development.

The main obstacle for their efficient use is not rather economic crisis, from which all the republics of the region suffer now, than from the existing interrelations between them. The reason of all this consists in the breakup of traditional ties and euphoria of independence of young states and well-known populism of politicians.

Therefore the top-priority objective today is overcoming all this and forming good neighborly mutually beneficial relations between the republics of the region.

CHALLENGES OF TRANSBOUNDARY WATER MANAGEMENT IN THE DANUBE RIVER BASIN

BENDOW, J.
International Commission for the Protection of the Danube River

THE DANUBE RIVER BASIN

The Danube River is 2780 km long and drains 817000 km² in the Central and Eastern part of the European Continent and flows from the Black Forest to the Black Sea. 83 millon people call the basin their home. The basin area includes all of Hungary; nearly all parts of Austria, Romania, Slovenia, Slovakia and Serbia and Montenegro; significant parts of Bosnia-Herzegovina, Bulgaria, Croatia, Czech Republic, Moldova and small parts of Germany and Ukraine. Very small areas can be found in Switzerland, Italy, Poland, Former Yugoslav Republic of Macedonia and Albania.

The Danube River Basin is not only the geographical catchment area of the second largest river of Europe, it also has played in the past and still plays today an important role as a cultural and historical center of political, social and economic development in Europe.

SOCIAL AND ECONOMIC FACTORS

An in depth analysis of the social and economic context of the different countries in the Danube River Basin is necessary to understand

H. Vogtmann and N. Dobretsov (eds.), Transboundary Water Resources: Strategies for Regional Security and Ecological Stability, 73–82.

the problems of cooperation and the efforts to be undertaken to achieve common regional and global goals.

The analysis of economic disparities shows a clear trend of a west – east decline of the GDP from the upstream countries like Germany and Austria, with about 28,000 US$ per capita and year (in 2002), to the downstream countries of which the Ukraine accounts for less than 5,000 US$ per capita and year.

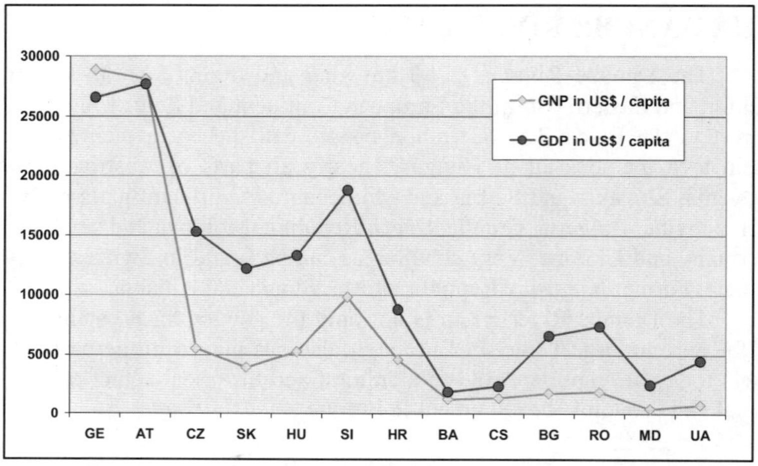

Out of the 83 million living in the Danube River Basin about 57% is living in urban areas. The share of population connected to public water supply varies from 29% in Moldova to 98% in Germany.

The share of population branched to public sewer system varies from 14% in Moldova to 89% in Germany. Based on the national projection figures, it can be anticipated that the population living in the Danube River Basin will by the year 2020 remain at its present level.

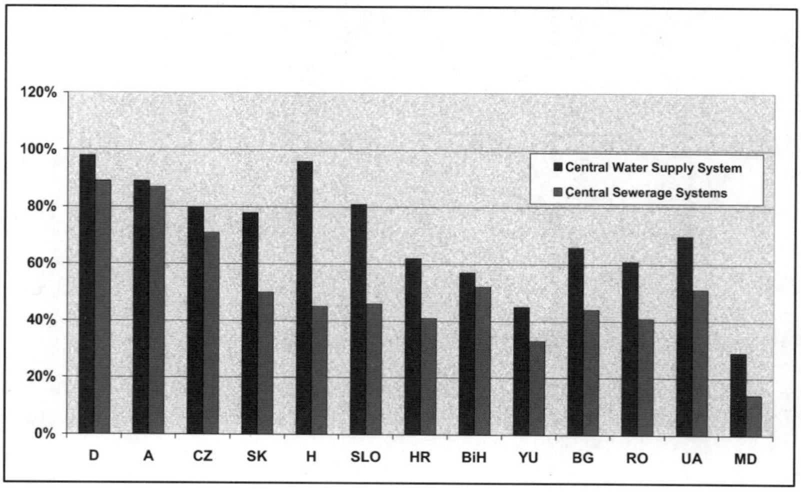

The middle and downstream Danube countries in transition are facing serious economic and financial problems to respond to the objectives of the Danube River Protection Convention and to implement measures for pollution reduction and for environmental protection as required for the accession to the European Union.

This analysis shows also that countries in transition need international support and makes evident the responsibilities of the international community to respond to regional and global concerns of environmental protection, with particular attention to:

- Restructuring and modernizing the legal and institutional framework and administrative systems;
- Establishing development policies and programmes as well as funding mechanisms in compliance with international standards of modern market economies;
- Initiating privatisation and establishing new links for international economic cooperation;
- Further harmonizing of national legislation with EU directives and standards.

OBJECTIVES AND OPERATIONAL MECHANISMS OF THE ICPDR

The Danube River Protection Convention is the legal frame for cooperation of the contracting parties to assure environmental protection of ground and surface waters and ecological resources in the Danube River Basin.

Out of 13 countries in the Danube River Basin, 12 and the European Commission have ratified the Danube River Protection Convention, which came into force in October 1998.

Objectives of the Danube River Protection Convention:

- Ensure sustainable and equitable water management;
- Conservation, improvement and rational use of surface waters and ground water;
- Control discharge of waste waters, inputs of nutrients and hazardous substances from point and non-point sources;
- Control floods and ice hazard;
- Control hazards originating from accidents (warning and preventive measures);
- Reduce pollution loads of the Black Sea from sources in the Danube river basin.

Contracting Parties to the Danube River Protection Convention

Are Germany, Austria, Czech Republic, Slovakia, Hungary, Slovenia, Croatia, Serbia-Montenegro, Bulgaria, Romania, Moldova,

Ukraine as well as the European Commission; Bosnia-Herzegovina holds observer status.

For countries holding a Danube basin area smaller than 2000km², the Danube River Protection Convention does not apply. However also these countries are cooperating with the ICPDR under the EU Water Framework Directive: Italy, Switzerland, Poland, Albania, and Former Yugoslav Republic of Macedonia.

The International Commission for the Protection of the River Danube

Is the coordinating body for the implementation of the Convention.

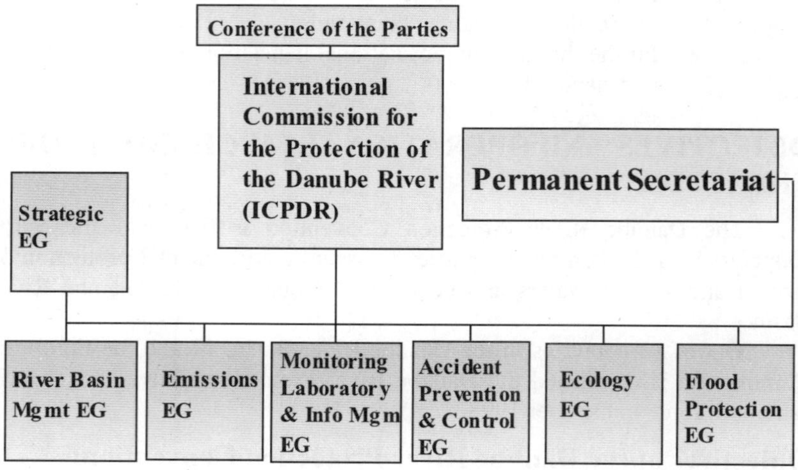

To facilitate cooperation in dealing with transboundary issues, specific Expert Groups have been created wit the following mandate;

The River Basin Management Expert Group (RBM EG)

Focuses on the implementation of the EU Water Framework Directive, in particular on the preparation of the River Danube Basin Management Plan. Upon agreement by the European Commission the measures foreseen in this plan will become legally binding for all EU Members States. From the States in the Danube Basin six (BG; CZ; HU; RO; SK; SI) are currently in the EU accession process.

The Ecological Expert Group (ECO EG)

Supports the activities of the ICPDR related to the conservation, restoration and sustainable management of the aquatic ecosystems and wetlands and the ecological requirements of the EU Water Framework Directive.

The Emission Expert Group (EMIS EG)

Is responsible for developing programmes to control and reduce pollution from point and diffuse sources (e.g. from municipalities, industry, agriculture). It facilitates exchange of information

The Monitoring, Laboratory and Information Management Expert Group (MLIM EG)

Is responsible for steering and evaluating the Trans-National Monitoring Network for water quality, for setting up programmes to improving the laboratory analytical quality assurance. It facilitates exchange of water quality and quantity information.

The Accident Prevention and Control Expert Group (APC EG)

Is responsible for steering and evaluating the effectiveness of the Accident Emergency Warning System. The system communicates messages among Contracting Parties about the emergency situations that may have a transboundary effect. Accident emergency prevention and control is the second main set of tasks, in particular, for developing tools and measures.

The Flood Protection Expert Group (FLOOD EG)

Is preparing a basin - wide action plan for sustainable flood protection and will then supervise its implementation.

The work of the Strategic Expert Group (S EG)

Aims at assisting ICPDR with specific advice on legal issues.
In addition to the main Expert Groups, Subgroups and Technical Working Groups have been created in the following fields:

Economic Subgroup:

Preparation of the economic analysis of water uses for the Danube River Basin Management Plan.

GIS Subgroup:

Development of Danube GIS and respective maps for the development of the Danube River Basin Management Plan.

Sava River Basin Management Working Group:

Development of the roof plan for an integrated river basin management plan on the Sava sub-basin level.

Danube Black Sea Joint Technical Working Group:

Development mechanisms to facilitate and enhance the cooperation between the ICPDR and the Black Sea Commission to reduce pollution in the Black Sea catchment.

THE IMPLEMENTATION OF THE EU WATER FRAMEWORK DIRECTIVE

The Danube River Basin is the second largest river basin of Europe and territories of 18 states. The area of the Danube River Basin district includes the coastal waters of Romania along the full length of its coastline as well as the Ukrainian coastal waters extending along the hydrological boundaries of the Danube river basin.

The implementation of the EU Water Framework Directive is a requirement of all member states major but also accession countries and other countries in the Danube River basin have given their engagement to implement the EU Water Framework Directive.

Key Elements of the EU WFD

- **Sets uniform standards in water policy throughout the European Union**
- **Requires cross border cooperation for the development of integrated and coordinated river basin management**
- **Stipulates a defined time-frame for the achievement of the good status of surface water and groundwater**
- **Introduces the economic analysis of water use in order to achieve the most cost-effective combination of measures in respect to water uses**
- **Includes public participation (stakeholders incl. NGOs) in the development of river basin management plans**

For the Danube River Basin the ICPDR will serve as the necessary platform for coordination.

Depending on the issue at hand three levels of coordination can be identified:

1. Danube River Basin level
2. Bilateral / sub-basin level
3. National level

The development of the Danube River Basin Management Plan is based on the EU guidance documents of the Common Implementation Strategy.

At the River Basin level the following tasks have been identified to be carried out under the guidance of the ICPDR:

- Strategy for development of River Basin Management Plan

- Delimitation of the Danube River Basin District (including the coastal waters of the Black Sea)
- Development of issue papers and preparatory studies on special Danube River Basin topics:
  - Economic analysis
  - Transboundary issues
  - Public participation, Danube GIS & mapping criteria
  - Typology & reference conditions of water bodies
  - Artificial and heavily modified water bodies,
  - Significant pressures and impacts
  - Effects from human activities on ground water
  - Register of protected areas (species and habitats)

The expected result is a complete River Basin Management Plan for the Danube River Basin, including a programme of measures

→ Part A: the "roof" for the Danube River Basin Management Plan (including all issues of basin - wide importance)

→ Part B: the national sub-unit plans (further information on the national level)

→ In addition, sub-river basin plans are foreseen to be developed in 2009

INTERNATIONAL COOPERATION

UNDP/GEF Danube Regional Project

The Actions:

- Creation of sustainable ecological conditions for land use and water management
- Capacity building and reinforcement of transboundary cooperation for the improvement of water quality and environmental standards in the Danube River Basin
- Strengthening of public involvement in environmental decision making and reinforcement of community actions for pollution reduction and protection of ecosystems
- Reinforcement of monitoring, evaluation and information systems for transboundary pollution control and nutrient reduction

UNDP/GEF support for implementation of the EU Water Framework Directive by a set of different activities concerning analysis and development of proposals for:

- Surface and groundwater

- Agricultural non point pollution sources
- Industrial pollution
- Land use and wetland rehabilitation
- Economic tools for water management
- Public participation and NGO development
- Communication and information
- GIS/cartography

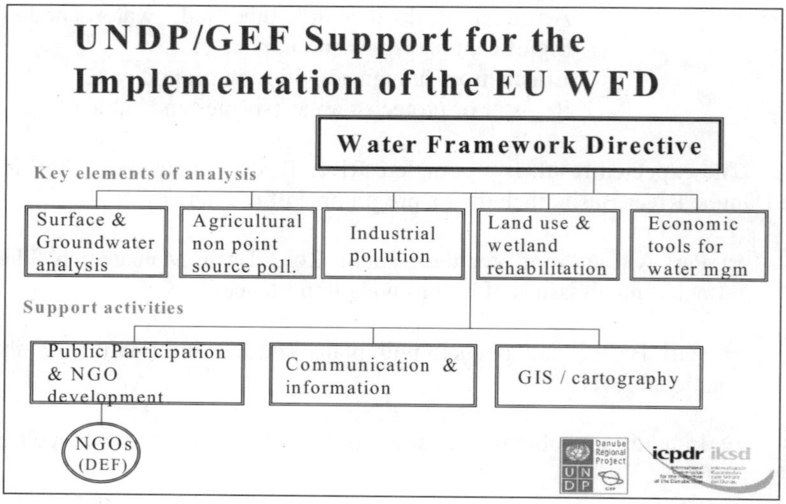

Cooperation with the Black Sea Commission:

The Cooperation with the Black Sea Commission is based on Memorandum of Understanding, signed in 2001 with the following goals:

- Long-term goal: to permit Black Sea ecosystems to recover to conditions observed in 1960s;
- Intermediate goal: to avoid nutrients load exceeding those in the mid of 1990s;
- Harmonization of standards;
- Assessment and reporting on ecological status and input loads;
- Adoption of strategies for pollution reduction whilst assure economic development;
- Analysis of results achieved by 2007 and review measures to achieve long term goal

ICPDR and Black Sea Commission are developing coordinating mechanisms for the implementation of the EU Water Framework Directive in the coastal areas of the Danube river basin district (Black Sea Coastal Waters)

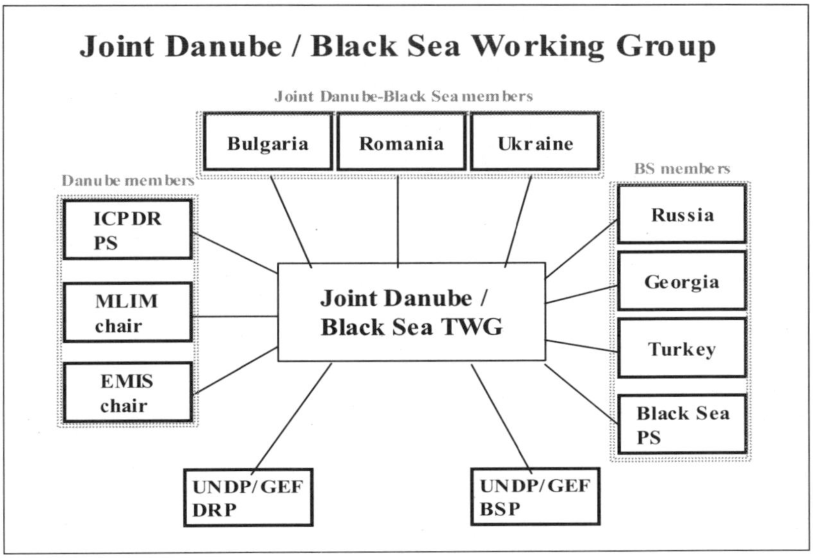

Danube Black Sea task Force (DABLAS):

Purpose of the DABLAS Task Force:

- Development of financing mechanisms
- Preparation and prioritization of investment projects
- Facilitation of cooperation between IFIs, bilateral donors and recipient countries (Donor Conferences)

The ICPDR has identified in the frame of the DABLAS initiative priority projects for municipal wastewater treatment facilities in all central and downstream Danube River Basin countries and has prepared a database, which facilitates the prioritisation of projects taking into account environmental and economic/financial criteria. In total 158 projects have been identified with a total investment of 2.6 billion € out of which 446 million € are secured.

The regional distribution of projects showing the state of development is indicated in the map.

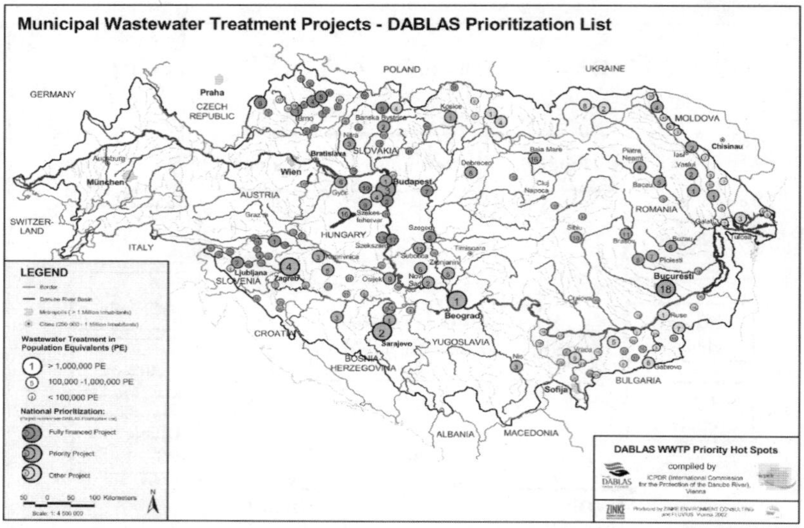

PPC/DABLAS Priority Project List

The final combined list of projects to be submitted to the international financing institutions from the Danube basin as well as the Black Sea area includes 34 projects with a need of more than 1.6 billion € (Danube basin: 21 projects, 1.5 billion € and Black Sea region: 13 projects, 150 million €).

These examples demonstrate that international cooperation is essential to assure sustainable management of international waters.

For more information, please visit: www.icpdr.org

TRANSBOUNDARY WATER PROBLEMS IN THE BASIN OF THE IRTYSH RIVER

VINOKUROV, YU. I., ZHERELINA, I.V. & V.I. ZANOSOVA
Institute for Water and Environmental Problems SB RAS

Irtysh River is the largest tributary of the Ob crossing the Asian continent from China to the Arctic Ocean. The Irtysh rises in China on the west slopes of Mongolian Altai and is called Cherny Irtysh till it empties into Lake Zaisan. The basin area makes up 1643 th.km^2, its length is 4248 km. Long-term average runoff volume in the Irtysh mouth is 89,3 km^3 (Fig. 1).

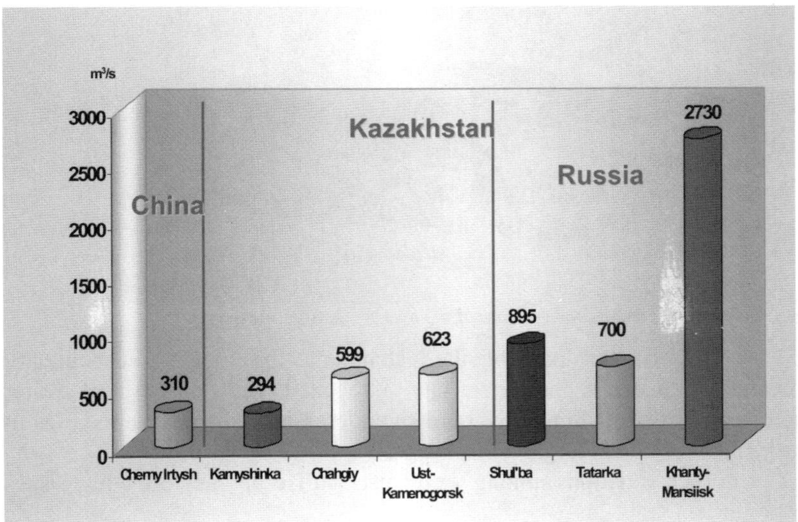

Fig. 1. Average long-term discharge in sections

The Irtysh basin is the international one partially joining the territories of three countries, namely Russia, Kazakhstan and China. The position of these countries in the catchment is different. Kazakhstan and China are in advantageous position since they are situated in the upper part of the basin. Russia occupies 45,4% of the basin area, however it has less winning geopolitical position since it is located in the middle and low part of the catchment (Fig. 2,3).

H. Vogtmann and N. Dobretsov (eds.), Transboundary Water Resources: Strategies for Regional Security and Ecological Stability, 83–91.

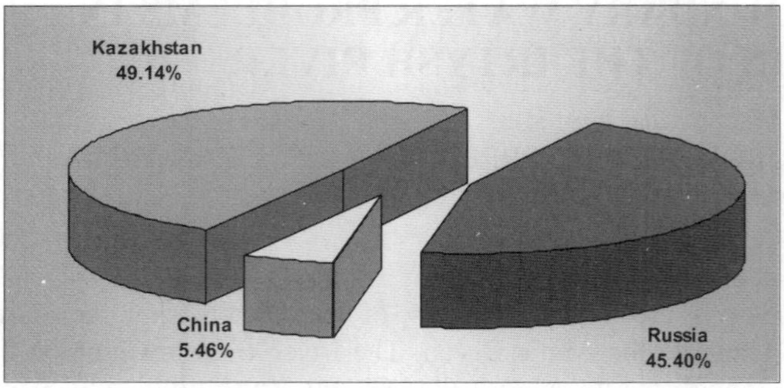

Fig.2. Relationship among catchment areas, %

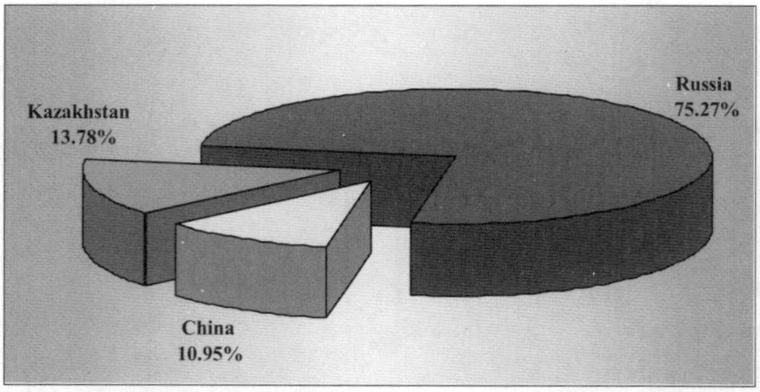

Fig. 3. Relationship among the volumes of the average long-term runoffs, %

In the Russian part of the catchment, the major transboundary problems of water management are associated with the use of water resources of Irtysh, Ishim and Tobol rivers crossing the state boundaries in Omsk, Tyumen', and Kurgan oblasts.

Conflict situations among water users arise on the following issues [1]:

- Coordination of water users' interests in the basin of transboundary river;
- Redistribution of water resources among the countries;
- Transboundary transport of pollutants;
- Interrelations under environmental damage elimination;
- Conservation of the unique water objects;
- Monitoring.

Let us consider in greater detail some points at issue.

Insufficient coordination in water management is the main reason of the problems occurred among the users of transboundary rivers. The most pressing conflicts arise on the problems of municipal drinking - water supply, navigation, fishery, recreation and hydropower engineering.

As a rule, water users' demands exceed real provision with water resources that causes the conflicts between water supply and hydropower engineering, hydropower engineering and fishery, fish industry and irrigation.

The conflict between hydropower engineering and fishery deals with the Irtysh regulation in Kazakhstan by Bukhtarminskoye and Shul'bunskoye reservoirs. Hydropower constructions hamper fish spawning due to low water during spawning and wintering migrations.

The conflict between fish industry and irrigation has resulted from the extensive development of river transport followed by transformation of river-bed to create favorable conditions for navigation, conducting dredging works and increase of inert building materials mining in the river-bed. These combined factors caused the deformation and slump of the Irtysh bed within Pavlodar town and the boundary between Kazakhstan and Russia.

The examples of negative aftereffects of water resources exploitation on the contiguous territories without coordination with the interested parties show that the solution of national or sector problems in one-sided capacity can bring harm to the interests of other parties and have an adverse effect on the basin ecosystem. Coordination of water users' interests in the basin of transboundary river is required.

Redistribution of water resources among the countries is a vivid example of an uncoordinated policy in water management exhibiting a strong negative influence on the total water stock. During last decades, the Irtysh flow has drastically reduced due to the construction and operation of reservoirs in Kazakhstan and extra-rate diversion flow as well (Table 1).

Name of reservoir	Volume of reservoir, km^3		Area of water surface, km^2	Types of use*
	Total	Effective		
Bukhtarminskoye	49,7	30,6	5490	PFNFiR*
Ust-Kamenogorskoye	0,65	0,17	87	IRFi
Shul'bins-koye	2,4	1,5	507	IRFi

Table 1

*Note: * W – water supply, I – Irrigation, T – timber rafting, F – flood control, R – recreation, F – fishery, N – navigation, P – power engineering*

The probability of aggravation of water-ecological situation in the catchment caused by realization of irrigation projects of China going to increase the offtake from Kara-Irtysh river from 0,45 km^3 to 1,5 km^3 /year (that is 20% of the river flow) is high.

The aftereffects of the project of diversion runoff from the Ob's runoff-part to Central Asia and Kazakhstan, which is the subject of wide speculation in Russia, are unexpected.

High regulation of the Irtysh runoff on the territory of Kazakhstan has led to the decrease of river runoff in Russia. Mainly it concerns the Bukhtarminskoye reservoir operating since 1960 and taking up 60% of the annual Irtysh flow. Since the river runoff has been regulated, the flood-land of Irtysh River in Omsk, a 500 km stretch (from the south border to Bol'sherechie) is not flooded as a result of river shoaling and extinction of 80% of flood-plain water bodies.

The problems in water supply and navigation are observed. Reduction in species composition and nutrient value of forage grass is found.

Transboundary transport of pollutants is one of the "hottest" problems. Russia is at a less advantage position in the Irtysh basin since it intakes the polluted water from China and, especially, from Kazakhstan where the following large industrial enterprises are situated:

- Lead-zinc and titanomagnesium plants (ust-Kamenogorsk town);
- The Irtysh and Leninogorsk polymetallic plants;
- East-Kazakhstan copper-chemical plant;
- "Aluminium of Kazakhstan" joint-stock company;
- The Aksus plant of ferroalloys;
- The Aksus thermal electric power plant;
- Petroleum processing and tractor plants (Pavlodar town).

The quality of river water entering Russia from Kazakhstan is estimated according to the data obtained on the section of Tatarka settlement on Irtysh River. Here, in 1998 the average concentration of ammonia nitrogen made up 2,3 of MPC, phenol – 3, copper – 10, iron – 4, zinc – 6, manganese – 11, oil products – 3 of MPC. River pollution by oil products is a result of a large number of damages at the production sites and during transportation. Annual damage caused by the entering pollutants makes up about 260 mln. Rubles [1].

Concerted actions of water users under environmental damage elimination are of significant importance. Emergency pollution of transboundary water can occur during the unit discharge of pollutants into Irtysh River and its tributaries. Unit discharges taking place in one country damage transboundary inner waters and water ecosystems located in the part of catchment in another country. It threatens the vital activity of population in all countries united by the Irtysh water system and hampers the successful operation of economic complex.

Interrelations between the parties, regulating the actions concerning the emergency situations in one of the contiguous territories, should be directed to transboundary water protection from pollution caused by hazardous activity, damages and natural disasters as well as to the effective response to emergency situations.

In doing so, responsibility for the pollution should be placed on the guilty party.

Conservation of water ecosystems of special natural, genetic, scientific, ecological-cultural, recreation and sanitary significance is a special link in transboundary water objects management. The sources of Irtysh River (Cherny Irtysh) found in China are included in Altai-Sayan ecosystem /ecological region. It is one of 238 global ecological hot spots marked by WWF as the prior ones for conservation of world biodiversity.

Economic activity within the catchment should be conducted on the base of the principle of water ecosystems' conservation. The establishment of the unified interstate system of water objects monitoring in the Irtysh basin is one of the main tools for its realization. However, there is a lack of a unified organization system determined by several reasons. One of it is the non-representation of the current observation net and the lack of the control stations situated on the states' frontiers. Besides, the lack of data and information exchange among the countries of the basin on water resources' state [2].

Thus, the problem of the development of an effective system for transboundary water objects management, use and protection is the main issue for all countries in the Irtysh basin.

The governments of the basin countries have repeatedly made attempts to find a solution for the problems of transboundary water use. In 1970-80 the issue on the development of "The Scheme of multiple use and conservation of the Irtysh water resources" was brought up at intergovernmental committees. In accordance with the Russian-Kazakhstan agreement (1992), the international committee on the problems of water resources use in water-scarce basins of Ishim, Tobol and Irtysh rivers was established. In 1996 the Russian-Kazakhstan committee worked out and adopted the protocols on joint use and protection of water resources, coordination of water economy activity in the Tobol and Ishim basins aimed at coordinated solution of the problems. The protocols include the volume of the runoffs, the procedure of information exchange and interactions under emergency situations. Besides, the committee elaborated and adopted the schedule of joint observations of transboundary water state, list of the indices under control, control intervals and methods for water samples analysis.

In 1995 the Board of "Siberian Agreement" Interregional Association rendered a decision to entrust the Executive Managerial Board and Coordination Council on Ecology with the development of the project of interstate basin agreement on cooperation between Russia and Kazakhstan in water resources management and conservation.

However, the developments haven't been realized completely, on grounds of political questions/ due to the lack of financing, political causes or the infringement of terms of joint work.

Nowadays international relations aimed at the solution of the problems of transboundary water use in the Irtysh basin have been resumed.

In May 2001 the fourth round of Kazakhstan-China consultations on the use of transboundary water took place in China. At the first stage the experts are expected to exchange information on the current situation in the Upper and Low Ob. Afterwards they will prepare the text of agreement on joint water use in the Irtysh basin.

In 2000-2002 the projects on the improvement of ecological situation and balanced management of water resources in the Irtysh basin were developed and have been implemented on the initiative of the World Bank, FFEM and DFID international foundations, and French companies ANTEA, SAFEGE, Office International de I'Eau with the participation of public organizations from Russia and the Republic of Kazakhstan. The largest ones are the French-Russian-Kazakhstan project "Transboundary management of water resources in the basin of Irtysh River" and the French-Kazakhstan project "Action program on the improvement of water quality in the Irtysh river basin". Notwithstanding the fact that the problems of water use and measurements on its improvement were worked through, they turned out to be poorly scientifically grounded and were not approved and supported by the government.

In 2002 within the framework of the Association of Academies of Science of Asia and under participation of IWEP SB RAS, the "Clean water – a step forward" program was elaborated. Also the international workshop on determination and establishment of priority of Irtysh river basin problems was held.

In June 2003, the International Conference on the problems of rivers in Ob-Irtysh basin was arranged [3]. The aim of the conference was to discuss the problems and measures on improvement of environmental and hydroeconomic situation in the Ob-Irtysh basin, to establish long-term collaboration among non-governmental, scientific and state institutions (involved in management and protection of rivers in Kazakhstan and Russia) and to work out the plan of joint activities on improvement of surface and ground water ecological state.

The participants adopted the following resolution:

- To establish on government level the Ob-Irtysh basin Council;
- To develop the project on interstate basin agreement on Ob and Irtysh rivers for water resources management and conservation;
- To work out the Interstate program on complex use, restoration and conservation of water resources in the Ob-Irtysh basin;
- To systematize and expand the monitoring network; to create the unified data bank
- To unify the rate of assessment of water quality in Irtysh river;
- To make an independent inventory of the sources of water objects pollution;
- To pass to ecologically safe technologies in industrial production;
- To establish a natural-ecological reserve in the delta of Cherny Irtysh river.

In the course of the conference the IWEP SB RAS researchers proposed the model on perfection of water use management in the Ob-

Irtysh basin approved by both countries and accepted for its further realization.

In line with international practice the decisions concerning interstate problems should be based on the international basin agreement signed and ratified by all countries situated on the catchment basin.

The agreement aimed an achievement of optimal conditions for water use meeting the interests of all countries that are situated within the basin and maintain high environment quality.

The agreement is based on the key principles required for further collaboration:

- The sovereignty of the state over natural resources on its territory;
- Responsibility of the states for the conservation of the basin environment;
- Balance between ecological human rights and economic development of hydroeconomic systems;
- Equal rights for transboundary resources use;
- Use of the territory without damage;
- Mutually beneficial cooperation in the solution of water use problems;
- Prevention of conflict situations;
- Mutual aid of the states in emergency situations;
- Regular exchange of information on the basin state;
- Peaceful settlement of ecological conflicts.
- Major functions of the Council:
- Coordination of the countries' activities in the use and conservation of transboundary water resources;
- Unification of water legislation, rates and standards in the field of water use and ecological safety;
- Development of interstate programs and activities on the use and conservation of transboundary water resources;
- Organization of international expertise of large hydroeconomic projects;
- Organization of water use monitoring in the basin;
- Development of a scientific and technical cooperation in water resources management and conservation, introduction of new technologies;
- Attraction of international and domestic foundations to support and protect the unique water objects;
- Control over the observance of terms of interstate agreement and implementation of the programs and activities approved.

The Interstate Basin Council is formed of representatives from federal bodies, regions' state authorities, and environmental non-governmental organizations. The decisions should be adopted by consensus since it allows the interests of all parties to be considered. The Council's work should be coordinated with the one made by other institutions charged with water use control and management at the interregional level in each country of the Ob-Irtysh basin (e.g. the Ob Basin Council in Russia).

Unfortunately scientific achievements in solving the problems of transboundary water resources remain practically unclaimed by branch managerial structures. Skilled scientists have to fulfil orders of some companies or institutions.

IWEP SB RAS is a specialized institute of the Siberian Branch of the Russian Academy of Sciences in the study of problems of nature management and water resources' state as well as of environment protection. Long-term research and assessment of the water ecosystems' state in the Ob-Irtysh basin are carried out here.

The concept of rational water use and the method for water use management in the basins of water objects of different hierarchical level were developed. It should be noted that the method mentioned above was used as a basis for the development of Concepts of State program on the use, restoration and conservation of water objects in the basin of Upper Ob river (2002-2010) and regional subprograms for the Russian Federation subjects in the basin of the Upper Ob ("Water in Russia – XXI century" (2003-2015) national program on further improvement and development of hydroeconomic complex of Russia) [5].

The Institute takes the lead in integration projects on fundamental research of SB RAS: "Ob-Irtysh basin system: formation, anthropogenic transformation, ecological state and strategy of water use". In 2003-2005 the Institute acts as a Principal Investigator of the "Global and regional transformation of water and chemical runoff in the Ob basin under the influence of natural and anthropogenic factors" project. It's an interdisciplinary integration project on SB RAS basic research. The investigations carried out within the framework of integration projects are the basis for the development of the water resources management system in the international. Irtysh River basin

At present the international project "Scientific bases for integrated management of water resources in natural-economic and ecological systems of transboundary basins using the principles of sustainable development (the basin of Irtysh River as a case study)" is being developed at the level of academies of sciences of Russia and Kazakhstan.

Outcomes expected:

- Principles, criteria and methods for the assessment of stability in water supply of natural-economic systems of transboundary basins;
- Current regularities in dynamics of the long-term and annual regime of the rivers in the Irtysh basin;
- Assessment of possibility and expediency of depth regulation of the Irtysh runoff and inter-basin redistribution of water resources;
- System of the ecological restrictions on anthropogenic transformations of aquatic system of the Irtysh basin;

- Proposals to interstate agreements (China-Kazakhstan-Russia) regarding joint use of transboundary water resources in the Irtysh basin;
- Assessment of possible balanced water supply of the natural-economic system of the Irtysh basin by criteria of sustainability.

The problems of Irtysh River could only be solved in the context of a general strategy for the Central Asia region development. The projects on the stable water supply in the Irtysh basin should be closely associated and coordinated with the program on transboundary biosphere territory establishment in Altai, "Our Common Home Altai" movement and other programs implemented in the Irtysh basin.

To achieve these goals, the support of international organizations and foundations (i.e. Regional Ecological Center of Central Asia, World Bank, UN Programs of development, Global Ecological Foundation, World Wild Foundation), which have rich working experience in this region, is called for.

REFERENCES

1. Water of Russia. Social-ecological water problems. Edited by A.M.Chernyaev. MNR..Institute for water resources 'complex use and conservation. – Yekaterinburg: Aqua Press publ., 2001. – P. 364.
2. Water of Russia. Water economic structure. Edited by A.M.Chernyaev. MNR. . Institute for water resources 'complex use and conservation. – Yekaterinburg: Aqua Press publ., 2001. – P. 428.
3. Proceedings of International Conference on the problems of rivers of Ob-Irtysh basin – Kazakhstan, Ust-Kamenogorsk, 2003. – P. 80.
4. Water problems at the new millennium threshold. Edited by M.G.Khublaryan. RAS. Institute for water problems. – M.: Nauka, 1999. – P. 347.
5. The concept of State Program on use, restoration and conservation of water objects in the Upper-Ob basin (2002-2010 гг.) / Yu.I. Vinokurov, I.V. Zherelina, A.Sh. Khabidov & oth. – Barnaul- Novosibirsk: Publ. Ludvinsky & K, 2002. – P. 24.

TRANSBOUNDARY WATER PROBLEMS IN THE KUR – ARAZ BASIN

MAMEDOV, R.M.[1], MANSIMOV, M.R.[2] & KH. R. ISMATOVA[3]
1:Institute of Geography, Azerbaijan National Academy of Sciences 2: Institute of Radiation problems, Azerbaijan National Academy of Sciences 3: Azerbaijan Aerospace Agency

Key Words: water resources, Kur, Araz, drainage basin, water consumption, pollutants, ecological condition, regional security, information model, delta, sea level.

ABSTRACT

Water resources of the rivers Kur and Araz

Water resources of the rivers Kur and Araz are stated in the first part of the report. The rivers Kur and Araz are transboundary for Caucasus countries as well as for Turkey and Iran to a certain extent. Kura is the largest river of the South Caucasus, with a longitude of 1515 kms, the drainage basin has an extension of 188 thousand км^2 and is situated in the territory of the 5 following states: Azerbaijan - 52,9 thousand km^2, Iran - 40 thousand km^2, Georgia - 36,4 thousand km^2, Armenia - 29,8 thousand km^2, Turkey - 28,9 thousand km^2. The largest influx of the river Kur is the river Araz with the drainage basin of 102000 km^2. Below the estuary of Araz, apart almost 210 kms the Kur has no other influx.

The long-term mean volume of water resources in the basin of the river Kur is 26,6 km^3. The total amount of water use in the countries of → the river- basin is about 23 km^3, that means 86 % of water resources are used for the needs of population and economics. For the last 60 years, a trend in the party of decreasing of liquid water content of the river. Kur can be observed.

The analysis shows that the structure of water consumption in the countries situated in the Kur river basin is approximately following: irrigation – 68%, heat-power engineering – 11,0%, industry – 6,9%, communal economy – 6,3%, agricultural water supply – 5,2%, forestry – 2,6%. Water consumption increase is observed in Azerbaijan for the last 30 years (7,0%), East Georgia (7,0%), Armenia (8,5%), Iran (4,1%).

The resources of river waters of Azerbaijan are 980 m^3/sec., or 30,9 км^3. The considerable portion of flow in Azerbaijan enters from contiguous territories by the transit rivers. The mean value of this influx on the liquid water content per year is 652 m^3/sec or 20,6 km^3.

Ecological conditions and regional security of the region are described in the second part of the report. The problem of rational consumption and protection of water resources is very actual for the region

H. Vogtmann and N. Dobretsov (eds.), Transboundary Water Resources: Strategies for Regional Security and Ecological Stability, 93–107.

countries. The last 25-30 years no serious fundamental investigation on the assessment of water quality forming and transformation was carried out.

The data of water-security offices of Azerbaijan, Armenia and Georgia in 1998 in the basin of the river Kur indicated about 453 million $км^3$ of sewages too little, including within the limits of Armenia 212 million km^3 (47 % from all volume), in Georgia - 229 million km^3 (51 %) and in Azerbaijan - 12 million km^3 (5 %). Unfortunately the pollution of the rivers and reservoirs is going on in territory of the Azerbaijan Republic.

Water supply of the city of Baku is also analysed in this part. Regretfully, it is necessary to mark that now, all kinds of international terrorism are appearing. And all countries of the world try to avoid any vulnerability by all possible versions and means. In this schedule the water facilities of large cities are mostly vulnerable, especially, if it is on-line provided from the open reservoirs or rivers. The city of Baku, half maintenance by water, which is derived from the river Kur, is in the most dangerous situation.

Information model for delta of the river Kur was worked out by us. The significant part of the model is to give a possibility to integrate the data of remote sensing, geographic and cartographical information to obtain the reliable information about the condition of the delta in different time periods and to forecast for the future The analysis of information for different periods showed that in the delta of the Kur took place great changes. At the present, the volume of the firm flow is not more than 10 million tones in a year. After the intensive rise of the level of the Caspian Sea begun in 1978, 1/4 of the delta was aggradated. New brand inside of the riverbed was created. The scheme distribution of sedimentations in estuary zone of the river Kura in the Caspian Sea is also shown on the data of aerospace image (1990 -1998 years).

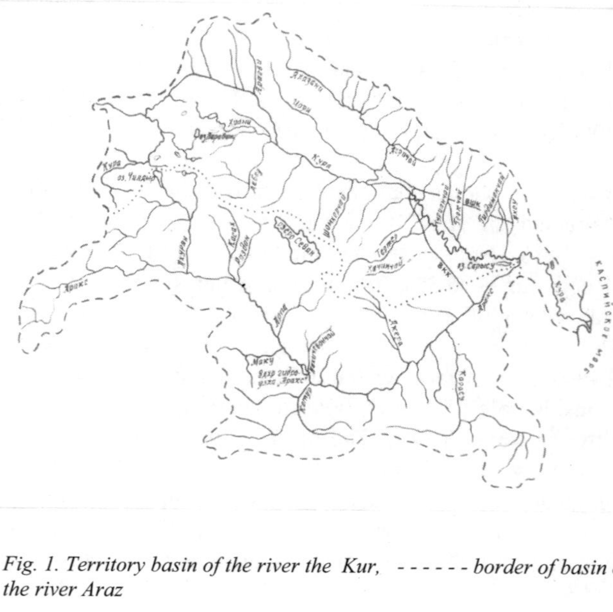

Fig. 1. Territory basin of the river the Kur, - - - - - - border of basin of the river Araz

WATER RESOURCES OF THE RIVERS KUR AND ARAZ

Nature and humanity → have always been in discrepant relations with each other, i.e. there was a time when nature threatened man, but now man threatens nature. Concerning the water resources consumption by people, the same is going on: uncontrolled exhaustion of natural water resources and disorderly feeding of sewage by people to the water objects, further the water shortage, the water pollution, unclean foodstuffs production, increase and spread of different diseases environment depletion and oppression. One of such regions of the world is the Caucasian region, especially its southern part, where vital activity of people and economic sectors fully depend on the quality and quantity of water resources. The water resources originate in the territory of one country, but they are used in the territory of others, i.e. side by side with natural factors, the anthropogenic impact on water resources make one country to some extent dependent from another one.

The countries of the South Caucasus are Azerbaijan, Georgia and Armenia. The rivers Kur and Araz are transboundary for these countries as well as for Turkey and Iran to a certain extent.

Kur is the largest river of the South Caucasus, its source is in Turkey (area Ardagan) at the altitude of 2740 m., and the estuary is in Azerbaijan on -27m. The length of the Kur is 1515 km^2, and the drainage basin 188 thousand km^2 (Fig. 1) and is arranged in territories of 5 states. It is arranged as follows: Azerbaijan - 52,9 thousand km^2, Iran - 40 thousand km^2, Georgia - 36,4 thousand km^2, Armenia - 29,8 thousand km^2, Turkey - 28,9 thousand km^2The longest part of the stream course flows through Georgia (37,7%), followed by Armenia (23,4%) and Azerbaijan (21,5%) as well as 13,6% in the territory of Turkey and 3,8% in Iran.

The basin of the river Kur covers 64 % of the territories of the countries of the Southern Caucasus. More than 65 % of the terrain of the basin of the river Kur (122,2 thousand km^2) is in an altitude of more than 500 meters above the sea level and introduces the area of a supply and transit of flow, and 35 % of the area are rearrangements and losses. The common slope of the river makes 2,03‰ (Rustamov S., Kashkay R., 1989).

The largest influx of the river Kur is the river Araz (fig. 1) with a drainage basin of about 102000 km^2, which is 54,2 % of the basin-area of the river Kur. The source of the Araz also is in Turkey at the altitude of 2990 m. (spine Bingel). Its length is 1072 kms, the mean slope 2,8‰. Below the estuary of Araz, for almost 210 kms, the Kur has no other influx.

The long-term average volume of water resources of the Kur-River basin is 26,6 km^3. The total amount of water use in the countries of the basin of the river is about 23 km^3, that means 86 % of water resources are used for the needs of population and economics. .20 million persons are living at the river basin. (Fig. 2). The enormous water use exhausts the rivers in the basin and decreases the flow at its estuary. As is seen from the figure. The decrease of water amount of the river Kur has been observed and became obviously during the last 60 years.

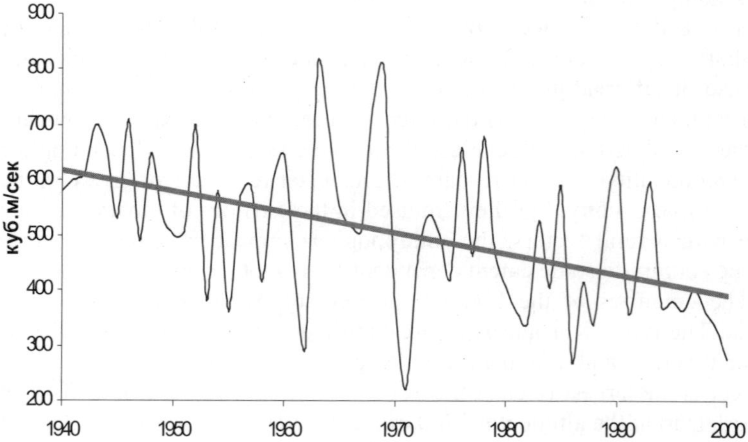

Fig. 2. Change of water content river Kur in cities Saqlyan.

The analysis of water consumption structure in countries situated in the Kur river basin is approximately as follows/ in the following way: irrigation – 68%, heat-power engineering – 11,0%, industry – 6,9%, communal economy – 6,3%, agricultural water-supply – 5,2%, forestry – 2,6%. Water consumption increase is observed in Azerbaijan for the last 30 years (7,0%), East Georgia (7,0%), Armenia (8,5%), and Iran (4,1%).

The existing condition has even more aggravated in the last years with the reduction of the natural flow in the river basins, in connection with climatic changes. In the last 6 years (1995-2001) the natural flowing diminished about 25-30%, in alpine regions the solid precipitations diminished and the seasonal snow line rose to 500-800 m. beyond usual. Thus, it is necessary to note that especially in 2003, the water amount of the Kur exceeded the long-term norm and flooded large territories in the southern flow.

This year, the Kur has caused many social and ecological problems in Azerbaijan: About 31 administrative regions with a population of 1,5 million people were affected.

In Azerbaijan, 980 m^3/sec., or 30,9 km^3 of water arise from the river resources. The considerable portion of flow in comes from contiguous territories by the transit rivers. The mean volume of this influx to the liquid water content per year makes 652 m^3/sec or 20,6 km^3. The stream course, directly reshaped within the limits of the country (local flow), is 328 m^3/sec or 10,3 km^3. As a whole, from the local resources of fluvial waters the portion of surface flow averages 58 %, and underground flow 42 %. The resources of fluvial waters of the country for a rainy year are 735 m^3/sec, or 23,2 km3, and for a year with little rain 228 m^3/sec or 7,2 km^3. It is necessary to note that 80 % of the counties water resources consist of the river Kur and its inflow, of which 70 % is reshaped in terrains of neighboring states. The water resources of the republic are limited extremely and contrasting to neighboring states, their specific weight per unit of territory and per capita accounts less according to Georgia 7.7 and 8.3 times, and to Armenia 2.2 and 1.7 times.

The main conclusion of The Second International Conference On "Climate and Water", which had taken place in Finland in 1998, and where I made a report (Мамедов et al., 1998), was concluded as following: In the future, the quantity of humidity will diminish in arid zones of the planet/earth/world and it will increase in damp regions. If we take into account that 60 % of the territory of Azerbaijan is part of the arid zone and already suffers an acute shortage of water today, it is not difficult to imagine the water problems which this country has to expect. Unfortunately, as mentioned above, the conclusion and the forecast already starts to slow down itself step-by-step concerning to the water amount of the rivers of Azerbaijan. As was marked above the natural flow has decreased about 25-30% during the last 6 years. So, the little water still decreases, the situation is aggravated because of anthropogenic and natural causes.

ECOLOGICAL CONDITION AND REGIONAL SECURITY.

The problem of rational use and protection of water resources is very actual for the region countries. Until 1990, the water use and water resources quality control was realized on the base of legislative acts of the former USSR and agreements between the USSR and Turkey and Iran. At the same time the state system of regular observations, water quality control and sewage and drainage waters cleaning and draining system were agreed. The scientific-research organizations of the country conducted the scientific-research and engineer-prospecting activities for rational use of water resources.

However, in the last 25-30 years not any seriously fundamental investigation on the assessment of water quality formation and transformation, and the natural facilities of self-cleaning and the renewal of water quality characteristics has been studied in order to develop the scientific-based measures on transboundary rivers quality management. Due to economic difficulties and the lack of programs and acts between countries on rational water resources use, and also the lack of understanding water capacity and quality formation regularities, water resources consumption and distribution at present time; the situation in the region is confusing.

Also the lack of a regional data collection and exchange system in the last 10-12 years makes it impossible for administrative bodies to make timely and objective decisions. There is no regional system of beforehand water capacity and quality prevention.

Azerbaijan faced such a situation in the period of 1999-2000 when it was impossible to plan the actions schedule of Azerbaijan big reservoirs timely: The levels of Mingechevir, Shamkir and Nakhchivan were below the dead capacity most part of the year. Accordingly, in lower reaches of the Kur and Araz rivers, the flowing was two times below the sanitary norm, but the degree of pollution on separate elements was above the utmost permissible norm. This process is going on up to present, and in spite of all kinds of cleanings, the drinking water for Mngechevir, Yevlakh, Ali-Bayramli, Baku and other towns is still polluted.

The water flow to Kur and Araz riverbeds below the sanitary norm leads to the increasing of underground waters flow, polluted with pesticides from agricultural lands to the riverbeds. Due to unprofitable flowing regulations, the intensive obstruction and silting of Kur and Araz riverbeds can be observed. The rivers lost their unique flora and fauna, which are relic for these places (rare species of fish, crayfish, forest plant attached to river-beds, etc.).

For the last years, the disposal of polluted sewage was realized only irregular, and all small and large river basins on the lower reaches of the river lost their natural self-cleaning-abilities The pollution of the separate rivers exceed the permissible concentration many times. The agricultural pesticides aggravated the situation. So, in 1992, the data of water-security offices of Azerbaijan, Armenia and Georgia concerning the basin of the river Kur indicated about 575 million km^3 of sewages too little, including within the limits of Armenia 300 million. km^3 (52 % from all volume), in Georgia - 250 million km^3 (43 %) and in Azerbaijan - 25 million km^3 (5 %). And in 1998, in connection with economical changes, these parameters were a little lowered again when common sewage disposal in the basin of the river Kura were 453 million km^3. For Armenia, this volume has gone down to 212 million km^3 (47 %) from all volume, Georgia down to 229 million km^3 (51%) and Azerbaijan down to 12 million km^3. As a result of this, the annual water flow of these rivers in their lower limits brings 7662 thousand tons of dissolved chemical combinations, 6060 thousand tons of suspended matters, 4-5 thousand tons of petroleum, 350 tons of phenol and up to 300 tons of compounds of metals. More than 60 % of these matters are on the share of the river Kur, 25 % on the share of Araz and the remaining 15 % - on a loop of the rivers Alazani, Iori, Akstafacay and Oxchuchay. The extremely unfavourable ecological situation was culminated by undercurrents of the rivers, first of all in the terrain of Azerbaijan, where already the flow from the terrain of Georgia was contaminated by industrial and municipal sewages and from the terrain of Armenia by cities of Yerevan, Kafan, Kadzaran, Alaverdi. 80 % of the population, 90 % of the agriculture and 100 % of the industry and other branches of economics use water resources from the river Kur.

Years	Phenols	Suspen-ded fragments	Petroleum	Heavy metals, Copper	Azote, ammonium
1970	0,005	675	0,06	0,012	5,2
1975	0,010	576	0,05	0,012	4,3
1980	0,011	834	0,025	0,012	8,3
1985	0,024	1050	0,11	0,010	4,8
1990	0,024	940	0,95	0,020	2,8
1995	0,045	1023	0,20	0,012	8,0
2000	0,036	994	0,18	0,028	5,0

Table 1. Concentration of some pollutants in the lower reaches of the river Kur, Mg/l

Unfortunately, the pollution of the rivers and reservoirs is continuing in the territory of the Azerbaijan Republic. The main contaminants of the rivers Kur and Araz within the limits of the republic are the industrial enterprises and municipal services of cites like Dashkesan, Ganja, Nakhchivan and Mingecevir. The sewages enter the flow after having passed one single sewage plant where only 30-40% from the total amount have been cleaned.

During the last time, the river Koshkarcay has been polluted daily by firms of the city Dashkesan, mainly an ore dressing factory, with another 46 thousand m^3 of sewages. This river runs into the river Kur almost at the tailpiece of the water storage basin at Mingecevir. Everyday about 300 thousand m^3 of sewages from the city Gandci is dumped in the river Gandcacay, which runs into Mingecevir water storage basin after 25 km. In the table 1 are given the concentrations of some pollutants in the undercurrent of the river Kur (The ministry of an ecology and natural resources of Azerbaijan, Annual report, 2000).

WATER SUPPLY OF THE CITY OF BAKU

Regretfully, it is necessary to mark that nowadays, we have to face all kinds of international terrorism. Therefore, all countries of the world try to avoid any vulnerability by every possible versions and means. In this schedule, the water facilities of large cities are the most vulnerable objects, specially, if they are addicted to the open reservoirs or rivers. The city of Baku, which obtains half its water supply from the river Kur, is in the most dangerous situation. 23,16 m^3/sec or 730374 thousand m^3/year of water enters the city, and more than 40 % derives from the rivers Kur and Araz, which, as seen from the above mentioned ecological parameters, already is dangerous enough for the health of the population, without thinking about terrorism In the tables 2 and 3 are given chemical compositions of the water of the river Kur at the main water collection construction of Kur water line for the city of Baku.

INFORMATION MODEL FOR THE DELTA OF THE RIVER KUR

The natural water volume changes in large intervals (as mentioned above, the water level in 2003 was very high) and additionally the anthropogenic factors in the last years have created great changes in the delta of the river Kur. But for all that, the delta of the river Kur is traditionally registered in the previous maps with three main arms (fig. 5a - map of 1982), which does not meet the actual situation. Aerial views of 1998 and 2000 demonstrate, that the bed of the river Kur, in the lowermost part of the delta (fig. 4, - the space image of 1998), has now changed considerably. Therefore, for observations and analysis of the factors influencing the changes of the bed of the Kur-delta, the information set model «Delta of the river Kur » was constructed by us. The significant part of the model is the possibility of integrating the data of remote sensing, the geographic and cartographical information, to obtain the reliable information about the conditions of the delta in different periods of time and to forecast for the future. The Methodology of the model is shown in the figure 3, its detailed definition is given in article (Mamedov et all., 2003.). All data are submitted in the base cartographic projection (UTM, WGS 84, zone 39). For this reason they are comparable in time-space aspect.

Date of taking of samples, 2000 y.	Azotes, mg/l			Si, mg/l	Oil prod., mg/l	Phenols	PAB
	NH_4^-				мг/л	NO_2^-	NO_3^-
16.02	0,004	0,017				0,109	7,16
11.04	0,0	2,73	0,18	0,015		0,035	5,1
22.06	0,0	1,86	0,0075	0,018	0,048	0,017	5,4
04.07	0,0	1,65	0,007	0,02	0,04	0,032	5,09
04.10	0,0	2,1	0,015	0,0015	0,02	0,015	5,09

Date of sample, 2000 y.	Oxidation pelerman. (MnO^-_4)	Degree of chromaticity	Coli-index, kl/l	Microbial unit. kl/l	Dissolved O_2, mg/l
16.02	3,10	12	43	72	9,5
11.04	4,29	12	460	95	7,67
22.06	3,4	12			
04.07	3,45	15	260	260	
04.10		12	117	117	7,4
14.11			1100	810	9,6

Table 2. Biogenic components and organic matters in main water intake of Kur water pipe for the city of Baku.

Date of taking of samples, 2000 y.	Be	Fe	Mn	Cu	Mo	Ni
16.02	0,00006	0,02	< 0,01	0,03	0,0092	0,0093
11.04						
22.06		0,02	0,0	0,0		
04.07		0,03	0,1	0,06		
14.10		0,01	0,0	0,02		
14.11	0,000065				0,0056	0,0058

Date of samples, 2000 y.	Pb	F	Cr	Zn	Cn	U	Total activity, β
16.02	0,00065	0,48	<0.01		0.0	0.006	2.37
11.04						0.108	2.1
22.06		0,47	0,004	0.001	0.0		
04.07		0,46	0,008	0.01			
14.10		0,45	0,0	0.0	0.001	0.062	2.31
14.11	0,00027						

Table 3. Pollutants of inorganic origin (mg/l) in main water intake of Kur water pipe for the city of Baku.

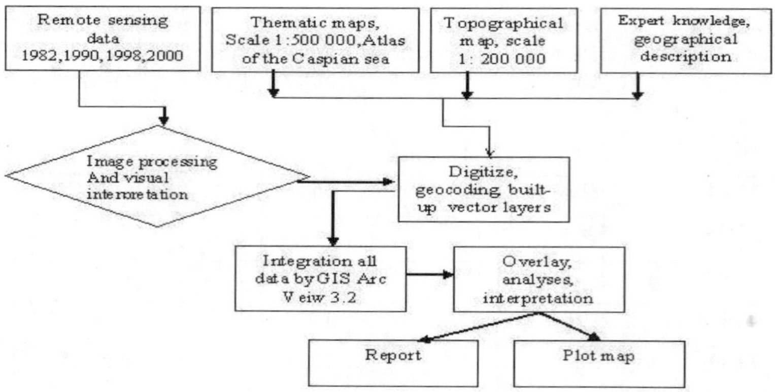

Fig. 3. Methodology of information model of the delta of the river Kur.

THE ANALYSIS OF THE INFORMATION

The lowermost region of the Kur in structural relation corresponds to the homonymous synclinal basin and the relief is characterized by the accumulative forms of relief. The subregion of the river Kur is completely accumulative formation. Now, the volume of a firm sink is no more than 10 million tons a year. After intensive rise of the Caspian Sea level begun in 1978, ¼ of the delta was affected by aggradation. In the last years, as a result of abrasion and flooding of low parts of the delta, its area was reduced almost twice. At the present, the process of abrasion is continuing in the front part. And half of the northern shore (Fig.4) is also subjected to abrasion. Except the deficit of drift deposits, this exerted influences the depth of the bottom (a lot). Naturally, the intensive raising of the sea level has aggravated dynamic processes.

In 1980, as a result of prolonged abrasion and the rise of the sea level about 0,5 m, a small strait was formed at the narrowest part of the spit (Fig. 5; a and b.).

At the present, the most part of the southern half of the fading spit was flooded by the sea. One third of the part of its area was kept/ is left. A small amount of material entering here from the abrasive area accumulates in the shallow-water zone. At the beginning, the width of the strait was equal to 2 km, but now the width of the strait is more than 10 km. The depth is 3,5 km (Fig. 5; a) and b).

The fact shown in the figures 5 a and b is a result of intensive rise of the sea level and wash of its shores by currents running through the strait into the Gizil-Agach gulf. Because of the weak current of sediments in the area of the spit-shore, the reunification of its two parts is unlikely the result of a 1-2 m sea- level fall. . In fig. 6 a and b are shown the modification of the delta and the site of concentration of sediments in the sea according to the pictures of 1990 and 1998.

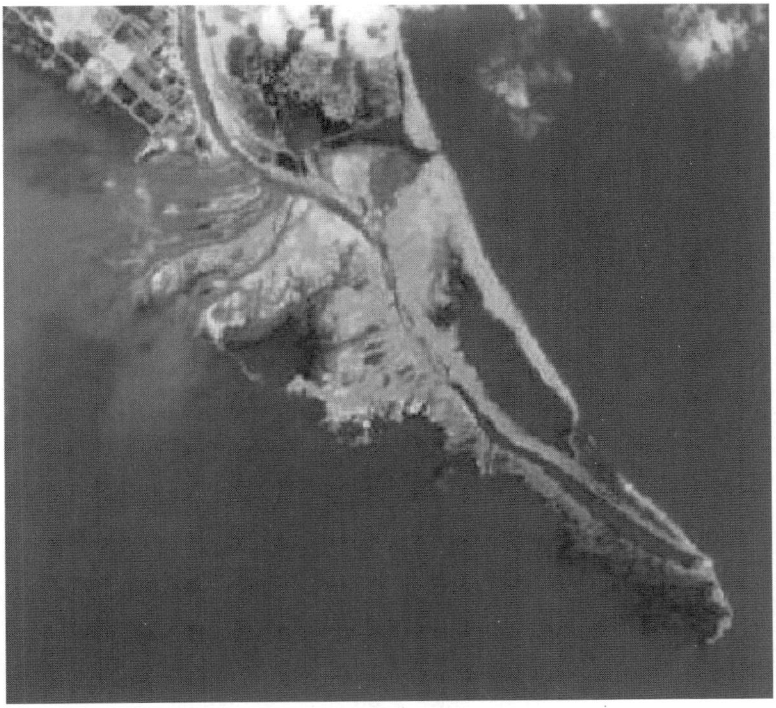

Fig. 4. Aerospace filming of a delta of the river Kur in 1998

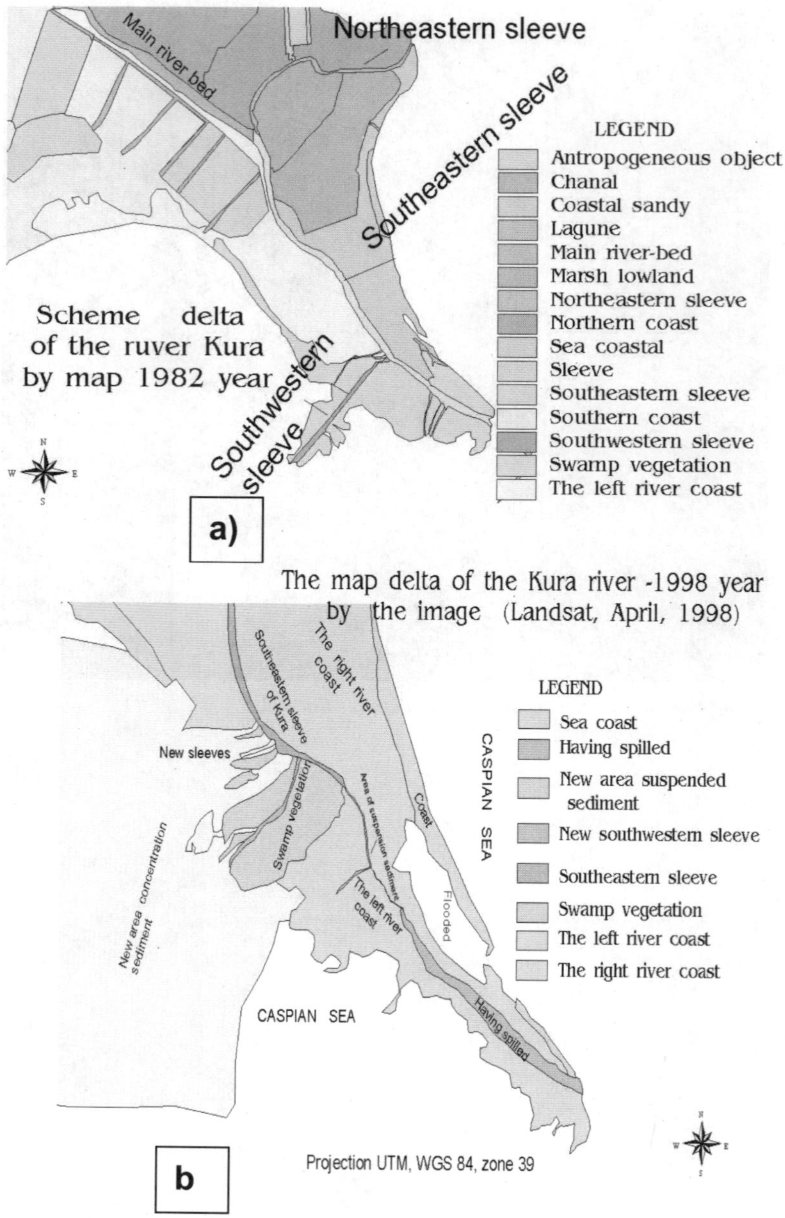

Fig. 5. The scheme of change of the water area of near-shore area of a delta of the river Kur: a) and b) layers accordingly imaging the information on a condition of basin of the delta of the river Kur on the 1982 and 1998.

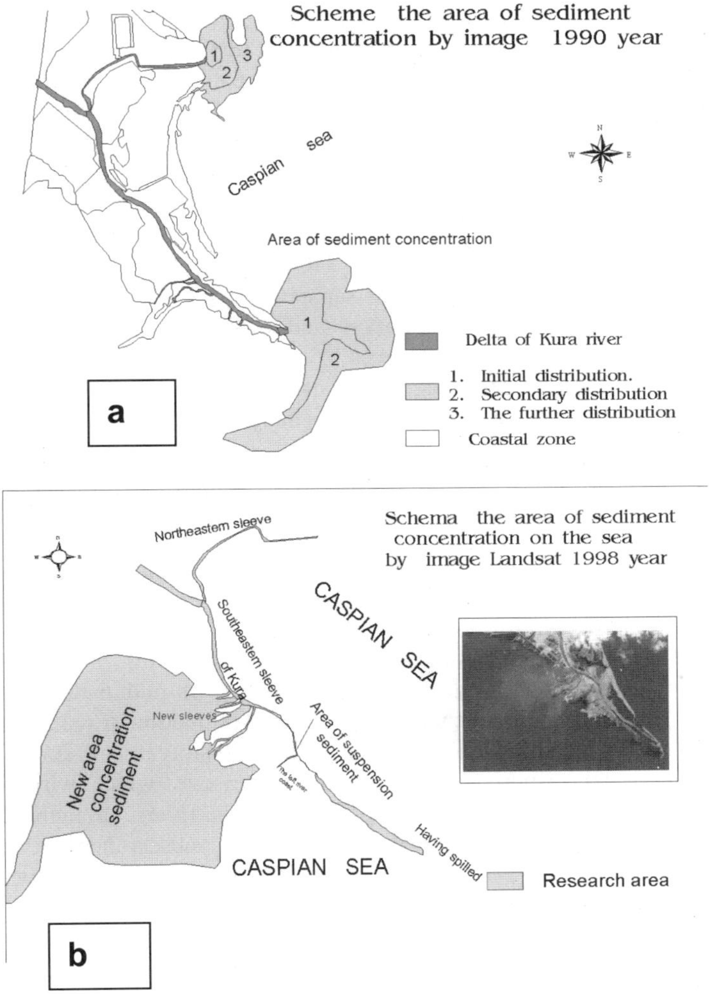

Fig. 6. The scheme of sedimentations distribution in estuary zone of the river Kur in the Caspian Sea; a) on the data of aerospace image of 1990 and b) on image of 1998.

CONCLUSIONS AND SUGGESTIONS

So, in order to evaluate completely the existing situation and to find a way out of the crisis, it is necessary to carry out a comprehensive work on the revealing of the sources of pollution, the regularities of formation and quality of pollution, transportation of pollutants along the length of rivers, the determination of the self-refining capability of rivers, reservoirs and all storage pools, to workout the scientific bases to prevent pollution, and the control and management of the water resources – quality of the basin of the Kur river.

It is impossible to carry out these works within the limits of one single country, even within the limits of countries of the region-Azerbaijan, Armenia and Georgia. Besides the unification of efforts of scientists and specialists of the corresponding organizations of these countries, it is necessary to take into account the advanced experience of developed countries and efforts of international organizations.

The following prime researches are suggested:

- Definition of the role of separate natural and anthropogenic factors for the formation and variability of quality and quantity of water resources;
- Estimation of the condition of the water quality in the river basins and tendency of their changes according to time and space;
- Definition of the most vulnerable segments of the basin and rivers which are subject to the intensive anthropogenic effect;
- Study of the transformation regularities and the turn into pollutants, their entry in the water environment and connected ecosystems;
- Definition of parameters of a river- self-refining
- Estimation of the effect of influence of pollutants on the ecosystem and health of the man, i.e. estimation of risk;
- Development of the action plan on quality management of water resources, their regulation and use, which will give measures to prevent the pollution of the rivers, strengthening of the water-self-refining capacity, observance of the sanitarian passing the reservoir, rational use of water resources.
- Special application of methods of remote researches, as the most perspective and all-embracing way of scientific collaboration.

REFERENCES:

1. S.G.Rustamov, R.M.Kashkay. Water resources of Azerbaijan, Baki, «Elm», 1989, 182 p.
2. R.M.Mamedov, L.I.Kulizade, Y.V.Hadiyev, Impact of climate anomalies on the level of Caspian Sea. The Second International Conference on "Climate and Water", Espoo, Finland, 1998, 17-20 August, p.972-980.
3. The Ministry of Ecology and Natural Resources of Azerbaijan, Annual report, 2000, page, 158.

4. 4. R.M. Mamedov, Kh. R. Ismatova, N. Mamedova. Using space borne data and GIS for research Azerbaijan coastal zone of the Caspian Sea. The Sixth International Conference on the Mediterranean Coastal Environment, October 2003, Ravenna, Italy (in print)

NATURE CONSERVATION AND SUSTAINABLE MANAGEMENT OF BIODIVERSITY

KINZELBACH, R.
Prof., Dep. Of Biodiversity Research, University of Rostock, Germany

Key words: Biodiversity, aquatic organisms, endemism, animal invasions, lake basins.

ABSTRACT

A peculiar aspect of inland water management especially in Central Asia

A brief introduction to the scientific field of biodiversity is given. Its value easily is underestimated. Reasons, general and economic ones, are discussed to save biodiversity. Therefore it should play a more important role in the future also in the field of water resources management.

The outstanding species- and population richness of aquatic ecosystems especially of Central Asia, harbouring a wealth of endemic species, is characterized. Lakes and rivers are not only a source of water for direct human use or a means to transport and clean wastewater. They have to be considered equally as the fundament for outstanding floras, faunas and habitats that affect indirectly human welfare and economy.

Aquatic biodiversity in Central Asia is endangered mainly by pollution, by loss of water by unsuitable irrigation projects and by direct or inadvertent transports of organisms across watersheds. This results in biological invasions, endangering endemic faunas and floras.

There are strategies to reconcile most of the traditional ways of use of water bodies and waterbeds (e. g. for industries, navigation, and irrigation) with the necessity to conserve water-related plant and animal life. This compatibility is examined. Proposals for advanced, integrated concepts of water management are made. They are determined to ensure conservation and continued sustainable use of biodiversity as a natural source for future economy and welfare.

Contents

H. Vogtmann and N. Dobretsov (eds.), Transboundary Water Resources: Strategies for Regional Security and Ecological Stability, 109–119.

4. Water management and biodiversity: An attempt of co-ordination

1. BIODIVERSITY: THE SCIENTIFIC FIELD

Life is diverse on all levels of complexity (molecular – cellular – organism – species – community). Of special interest is the diversity of species including their ecological niches. The value of species diversity and richness for the ecosystem and human being is stated by the convention on biodiversity (Convention on Biodiversity, Rio 1992) and by subsequent international agreements. Biodiversity is considered as a natural heritage and wealth. Its protection, connected with sustainable use, is one of the challenges of the 21st century.

The number only of living species (bacteria, fungi, plants, animals) which are scientifically described and registered at time is only for animals 1.85 Million; up to 20 Million animal species worldwide are expected still to be discovered.

Five global "Hot Spots" of outstanding species richness are identified, e. g. South East Asia, Madagascar, Ivory Coast, Central America. This frequently published assessment is based on plants and terrestrial animal life only. It neglects the facts

- That even small freshwater streams in the Northern hemisphere (Nearctic, Palaearctic) may harbour up to 800 species in only 10m of their course.
- That in the water bodies especially of the Northern hemisphere a highly dynamic evolution takes place, sometimes within decades, producing a wealth of infraspecific units, as subspecies, morphs, local populations etc., different from lake to lake and from watercourse to watercourse. So, each single water has its individual biodiversity pattern.
- That there is a tremendous degree of endemism especially among aquatic animals (vulnerable species restricted to very limited areas), due to their isolation in river and lake basins respective parts of them. Only to remember the largest inland basins of Central Asia as Lake Baikal, Lake Hubsugul, Lake Balkhash, and the Caspi-Aralian Basin: They harbour thousands of endemic species. (KOZHOV 1972, KOZHOVA & IZMEST'EVA 1998, KOSAREV, A. N. & E. A. YABLONSKAYA 1994, LÉTOLLE, R. & M. MAINGUET 1996, ILKIN, B. N. et al (1967).
- That biodiversity comprises not only the infraspecific and species levels but includes the diversity of habitats. This allows referring briefly to both, species and habitats together, as "nature".

Biodiversity research covers:

- Origin and structure of biodiversity, by the methods of systematics, including genetics, phylogenetics, geophyletics and taxonomy.
- Patterns and causes of the non-homogenous distribution of life concerning quality (species richness and species availability) and quantity (biomass production).

Biodiversity research is:

- The scientific fundament for protection and for sustainable use of plant, animal and microbial life by man.

2. FRESHWATER AQUATIC BIODIVERSITY IS ENDANGERED

2.1 General view

Generally River and wetland management is another term for an increased and intensified use of natural resources including biodiversity. Does human economy really threaten ecology?

Not necessarily (fig. 1), since there is a general identity of the ecosystem and the economic system. Economy is a mere subsystem of ecology, as man is part of the nature. Both correspond in the modes of input, in the central process of adaptation (including production, consumption, and destruction) and in loss of energy. In the ecological "black box" the process of species activity and adaptation takes place, depending on their identity and numerical relations. The same happens on economy's side considering specified methods of economic production, consumption and recycling. So, where is the problem?

There are problems:

- Intense population growth (the human success story) (fig. 2)
- Increase of mode and amount of consumption, of activity, of motility by man
- Still wrong or crude methods of production or management

Ad 1: Little can be achieved in short terms.

Ad 2: Little can be done, even should be done, since it would not contribute to solve any economic or social conflicts at time.

Ad 3: This presently is the only promising field for successful intervention. An improvement of water management is feasible. This fits into the frame of the ongoing workshop.

The viewpoint of the ongoing historical process of conversion leads to similar results (fig. 3). Man continues changing pristine nature continuously into an artificial habitat suitable to his demands, sometimes fortunately into "nature of second hand". Meanwhile nearly no spot is left on the globe where human influence is not traceable.

This process cannot be interrupted. Also here little can be achieved in short terms, and also from this viewpoint interference would not contribute to solve problems at time. The only successful type of intervention is to care for smoothening and moderating the transformation process, and for improvement of its management. This again describes the aim of this workshop.

All kinds of use of nature possibly threaten the richness of species: Their quality and quantity, globally, regionally, totally or partly. Traditionally this was accepted. Europe changed its pristine nature considerably in history. The process of re-thinking the relation to nature started only under the pressure of a decreasing quality of life. It was triggered by intoxicants in food, in the potable water, in the air; by noise, by traffic jams; by impoverished landscapes, crowded beaches…etc.

As fundamental reasons to take care for nature, always including biodiversity, are accepted:

- The wish for continued use: a product of sheer utilitarism (SINGER 1979)
- The desire to enjoy the diversity of life, the feeling to love or to miss the manifold animals and plants – a higher grade of utilitarism (SINGER 1979)
- Some people may add the argument of responsibility towards creation, which is **not** supported by the religions of the book. The biblical recommendation being: And God blessed them, saying, "Increase and multiply and fill the earth and **subdue** it…" (Genesis 1/28). There is no commandment to protect anything, except human existence and welfare. Again a utilitarian approach.

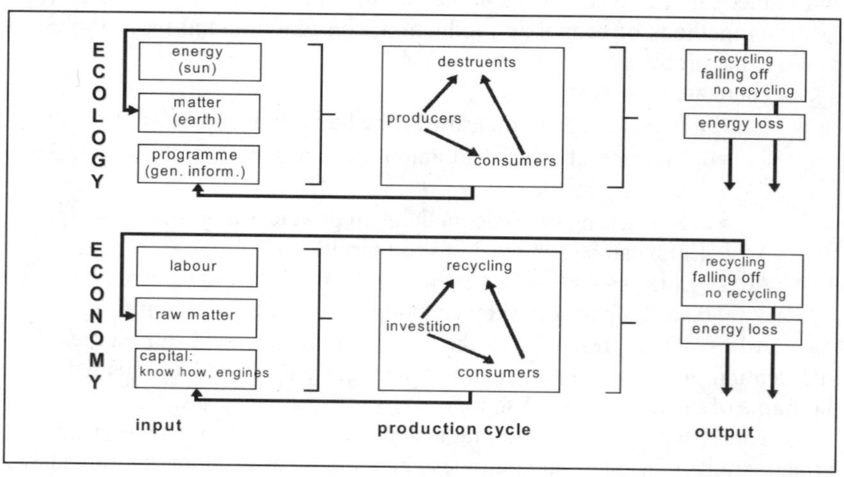

Fig. 1. The corresponding pathways of the global ecosystem and the human economy as its subsystem (KINZELBACH 1989, 1992).

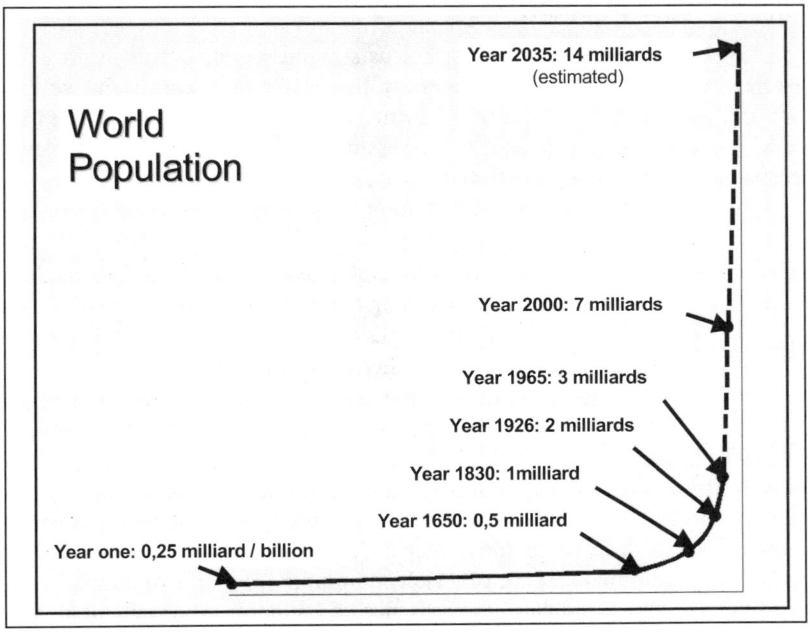

Fig. 2. Global population growth (KINZELBACH 1989).

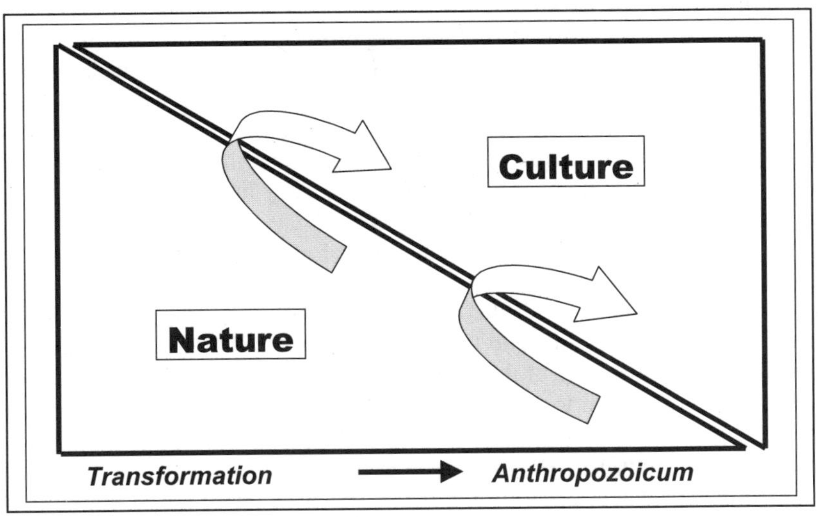

Fig. 3. Conversion: Man involves "nature" (the self-grown) more and more into his specific realm, the "culture" (literally agriculture, in figurative sense all human activity), which is characterized by his influence. This results in a state and an age of human domination.

2.2. Damages in detail

A widely spread attitude is: Let us sin and regret. Let us destroy and restore. Lets gain a flowering economy first, later on, being rich, we may care and pay for the restoration of former damages. This strategy does not work concerning biodiversity. Assessment of polluted or damaged ecosystems shows striking differences:

- Restoration or restitution is feasible in short-terms (years) concerning water pollution
- Restoration is possible, not equally easy, only in medium terms (decades) concerning soil pollution
- Restoration is difficult, possible only long-termed (centuries) concerning air or atmospheric pollution
- Restoration is impossible concerning life. Extinct species or populations and their unique genetic codes are irreplaceable.

Therefore the protection of biodiversity deserves the highest grade of priority. It includes anyway a protection of its environmental conditions and habitats: air, soil and water. Endemites, extremely vulnerable being restricted to limited areas, deserve the most intense degree of protection.

Vice versa the presence of a rich community of plants or animals or of a specialized animal indicates the presence of a suitable environment also for man. This is the methodical base of the bio monitoring.

A special type of damage for isolated aquatic faunas and floras, which are rich in endemics, is the exchange of species. This had a long history in the time of the Soviet Union, when deliberately edible fish, crayfish, and mussels, including also their food organisms as worms, snails, small crustaceans, were displaced all over water bodies and catchments of this wide territory in both, Europe and Asia. Imported species replaced local ones. They altered considerably the food chains and the basic relations in the respective ecosystems, as flow of matter, energy and information. Economic results were usually limited in yield and time. They mostly stayed far below expectations and prognoses

Beside this deliberately initiated faunal and floral exchange artificial connections of waterways caused unwanted biological invasions. This process started earlier, with the channel systems connecting the Black Sea watershed with the drainage to the Baltic Sea from 1790 on. One of the first species taking advantage of it was the zebra mussel (*Dreissena polymorpha*), which was considered to be a mayor impact to aquatic ecosystems from the beginning on (1844 Upper Rhine Germany, 1990 Great Lakes USA) (KINZELBACH 1992).

Planned water connections to the semiarid regions of Central Asia with the local lake basins containing specialised endemic faunas, may be devastating for biodiversity. By open canals or by pipelines aquatic plants and animals or their minute stages of transport or development will be dislocated and potentially they threaten as invaders the ecosystems of their destination. Suitable measures to prevent or control such transportations are urgently needed and should be taken into consideration for each water transport project.

3. THE VALUE OF PLANT AND ANIMAL SPECIES

Biodiversity is an economic factor. Goals and benefits of biodiversity conservation are generally accepted. An important viewpoint is:

- The estimation of cash value of biodiversity is necessary to include it into economic calculations. For this purpose case studies are necessary by regions (for each river basin), and by species.

An assessment of the **general** value is comparatively easy for forestry, wildlife etc on the terrestrial side (e.g. DUMOVA 2002, ATUTOV 2002). For these parts of the ecosystem traditional instruments of evaluation are available. For the aquatic biodiversity the value partly is less obvious, partly suitable methods still have to be developed or improved. Up to now two main types of use are described:

- **Direct** economic use by fisheries. This may include advanced types of aquaculture and the direct use of the flood plains by food gathering, cutting willows or reeds, hunting and by special types of agriculture.
- **Indirect** economic use by long-distance ecotourism, by recreation areas for cities in the neighbourhood, by the process of self-purification of intact ecosystems. The purification of organically polluted wastewater depends on biodiversity (mainly bacteria, but also algae, protozoans, nematods, rotifers, insect larvae etc). Also sewage plants are artificial ecosystems that instrumentalise also bacteria, protozoans, fungi and animals.

The **actual** value can be determined

- For selected single species by the evaluation of the potential income by fishing, fish farming or by assessment of the rarity (cf. Red Data Book for the Russian Federation) or by the functional value as part of the ecosystem (e.g. the value of amphipods after a chemistry accident in the Rhine was calculated 0.25 C per individual, based on their role as fish-food and on their participation in the process of self-purification of the polluted ecosystem) (KINZELBACH 1987).
- For an intact, attractive ecosystem by the number of paying visitors in a region, and the increase of quality of life.
- By calculation of the potential expenses to compensate, to replace or to restore losses of biodiversity. They may be extreme by re-stocking gaps by use of genetic reserves from neighbouring populations. Or, in many cases restoration is impossible (see above).

The restoration of the quality of the water body of the navigable part e.g. of the Rhine in Germany, approximately 800 km long, for **one single** point on the scale of the standard saprobic index, including the reduction of some top toxicants between 1970 and 1990, caused expenses of approx. 25 billion €. In this amount is **not** included the restoration of river banks and river beds, which, in a future semi-natural state, could serve as nearly 1.600 km of recreational beaches and areas for a densely inhabited region. This proves that it is wiser and by far less costly to save an ecosystem in advance than to restore it later – which for biodiversity, as presented, is frequently impossible (see above).

Examples for extremely valuable local species on the brink of extinction are:

- Two species of the Great Sturgeon (*Acipenser (Huso) huso* in the Pontocaspian basin; *Acipenser (Huso) dauricus* in the Amur river). The stocks of this largest fresh water fish are overexploited.
- Three endemic species of the sturgeon genus *Scaphirhynchus* in lower Syr-Darja and Amu-Darja, affluents to the Aral Lake. Their stocks are lost – no records within the last five years – due to chemical pollution, to lack of fresh water and to overexploitation. They are – or were – possibly very suitable for fish farming, allowing to combine sustainable economic use with protection of almost extinct species.

The message is: It pays out to protect life as a whole and in detail. It pays out to include it into concepts of sustainable development.

(4) WATER MANAGEMENT AND BIODIVERSITY: AN ATTEMPT OF CO-ORDINATION (RECONCILIATION)

Waterbed and water-body are well-separated parts of an aquatic ecosystem, but interdependent. It may be favourable e.g. for a fish species to find a clean water body; it will anyway get extinct if its spawning places (as gravel beds, plants) disappear; or if the food chain it relies on (e.g. insect larvae, water snails) is destroyed by covering the riverbed by concrete or stone blocks.

Water-bodies and waterbeds are used in several ways. A crosscheck of their compatibility shows that all modes of use are already or may possibly be made compatible, with the goal to protect biodiversity, with one exception: water pollution either by suspended or dissolved chemical loads, by radioactive or organic pollutants. In short terms and in terms of private economy it is an advantage to get rid of polluted water by rivers or lakes or the sea as transporting, diluting or cleaning systems. In long terms and in terms of political economy it is damaging.

Biodiversity in cases is a source of problems for the human population, causing or carrying water borne diseases, e.g. by bacteria, fungi, protozoans, helminth worms, and vector insects (e. g. mosquitoes). This is an additional reason to include the study of biodiversity into all current and future water management plans.

4. CONCLUSIONS

- Biodiversity conservation deserves highest priority
- Conservation can be integrated more or less into all concepts and plans of water management – except the discharge and transportation of intoxicated and/or organically polluted water
- Conservation is necessary as biodiversity is a natural resource
- Biodiversity pays out, it is a source of income
- Its use should be sustainable: Immediate and/or total exploitation should be replaced by concepts of stable long-term use
- Methods and expert-knowledge are available. They should be applied to regional and/or local projects beyond the present use and standard.

Tab. 1. Human use of biodiversity and its value:

Functionality of biological systems
Biogeochemical cycles; symbioses; organism complex. – Primary production including crop plants, domestic animals. – Raw materials by plants, animals. – Biological crop pest fighting. – Fertilisation of plants. – Biological sinks, filters, elimination of toxins. – Building up soil and humus. – Monitoring, bio-indication.

Source of biochemical information
Potential of natural adaptation. – New breeds and sorts; pest-resistent lines. – Pharmacology (bio-chemicals, genes).

Source of knowledge in science
Discovering new potentially useful species. – Bionics, eco-technology, energy saving technologies. – Basic understanding of biological and ecological systems.

Recreation, home, aesthetics (psychical wellness)
Phenotypic richness. – Richness of landscapes. – Richness of sensual impressions, of forms, colours, movement patterns, behaviour. – Intimacy, remembering. – Outdoor life, ecotourism.

Tab. 1. Human use of biodiversity and its value.

1 Potable Water
2 Shipway
3 Waste Water Transport
4 Fishery, Aquaculture
5 Cooling Water
6 Irrigation
7 Power
8 Recreation, Tourism
9 Protection of Biodiversity

Compatibility can be achieved: completely (dark blue), partly (blue), none (red).

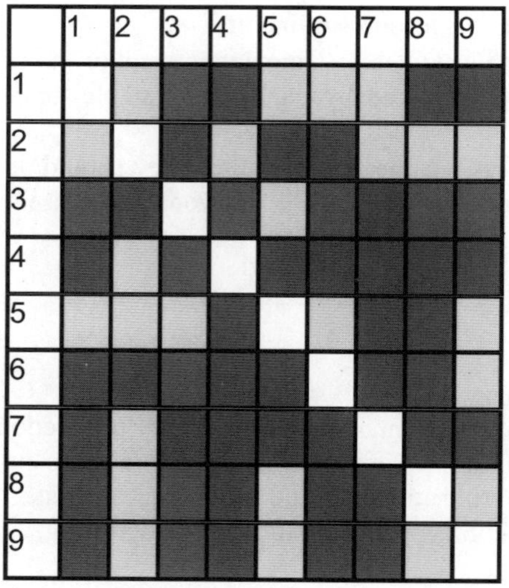

Tab. 2. Compatibility of different modes of use of waters. Most modes of use are already or may be made compatible with the protection of biodiversity: except waste water discharge/transportation.

REFERENCES (SELECTION)

1. ATUTOV (2002): Ecology and economy estimation of conservation and business use of biodiversity resources of riparian areas. - In: DUMOVA, I. I. (ed.) Irkutsk.
2. DUMOVA, I. I. (ed.) (1002): Nastolnaja kniga po ekonomike sochranjenija biorasnoobrasija Baikalskogo regiona. - 89 pp., Irkutsk.
3. ILKIN, B. N. (Ed.) with BIRSHTEINA, Y., L. VINOGRADOVA, N. N. KONDAKOVA, M. S. KUN, T. V. ACTAKHOWOI, N. N. ROMANOVOI (1967 pp): Atlas bezpozvonochnykh Kaspiiskogo Morya (Atlas of the Invertebrates of the Caspian Sea). - 416 pp., Moskwa.

4. KARPEVITE, A. F. (1975): Teorija i praktika akklimatizacii vodnykh organizmov. - 388 p., Moskva (Pishtshevaja proyshlennosti).

5. KINZELBACH, R. (1987): Die Tierwelt im Rhein nach dem November 1986. - Natur und Landschaft **62** (12): 521-526, Stuttgart.

6. KINZELBACH, R. (1989): Ökologie - Naturschutz - Umweltschutz. - 180 S., Darmstadt (Wiss. Buchges.).

7. KINZELBACH, R. (1992): The main features of the phylogeny and dispersal of the Zebra Mussel *Dreissena polymorpha*. - In: The Zebra Mussel *Dreissena polymorpha*. Limnologie aktuell **4**: 5-17, Stuttgart, New York (G. Fischer).

8. KINZELBACH, R. (1999): 2.7 Das Aufbrechen der ökologischen Frage. - S. 135-151. In: KORFF (ed.), Handbuch der Wirtschaftsethik. München.

9. Krasnaja kniga Rossijskoj Federacii (Zhivotnye) (2001). 860 p., Moskva (Asti Astrel).

10. KOPTYUG, V. A. & M. UPPENBRINK (1996): Sustainable development of the Lake Baikal Region. - 372 p., Berlin, Heidelberg, New York (Springer).

11. KOSAREV, A. N. & E. A. YABLONSKAYA (1994): The Caspian Sea. - 259 p., The Hague (SPB Academic Publishing).

12. KOZHOV, M. (1972): Lake Baikal and its life. - 344 p., Den Haag (W. Junk)

13. KOZHOVA, O. M. & L. R. IZMEST'EVA (1998): Lake Baikal evolution and biodiversity. – 447 p., Leiden (Backhuys Publishers).

14. LETOLLE, R. & M. MAINGUET (1996): Der Aralsee. Eine ökologische Katastrophe. – 517 p., Berlin (Springer Verlag).

15. Red data book Russia.

16. RUSSIAN ACADEMY OF SCIENCE / MINISTRY OF NATURAL RESOURCES OF THE RUSSIAN FEDERATION (2001): National Strategy of biodiversity conservation in Russia. – Moscow.

17. SINGER, P. (1979): Practical Ethics.

SUSTAINABLE WATER MANAGEMENT IN EUROPE – THE WATER FRAMEWORK DIRECTIVE

JEKEL, H.
Federal Ministry for the Environment, Nature Conservation and Nuclear Safety, Germany

ABSTRACT

Directive 2000/60/EC of the European Parliament and the Council of 23 October 2000 establishing a framework for Community action in the field of water policy (Water Framework Directive (WFD)) entered into force on 22 December 2000. The Directive as one of the most substantial pieces of EU water legislation combines the until then rather fragmented EU water law (large number of directives dealing only with special aspects of water management like waste water, dangerous substances, drinking water etc.) in order to ensure sustainable water management.

EU Member States have to define river basin districts as the main units for all management actions. For example the Rhine or the Danube with all their tributaries, associated groundwaters and coastal waters are two of the biggest international river basin districts in Europe. The river basin approach requires transboundary water management and therefore a lot of cooperation and consultation among the EU Member States. Although there have been several bilateral agreements and international river basin commissions in Europe until now water management happened mainly on national level. To fulfill the WFD requirements cooperation has to be intensified and new structures have to be established in the future.

The WFD aims at achieving good status of all water bodies (good ecological and chemical status of surface water bodies, good chemical and quantitative status of groundwater bodies) in a river basin district until December 2015 (with possible prolongations until December 2027). Good status is defined by ecological, chemical and quantitative criteria, which are described in detail in the annexes of the WFD. Ecological criteria, which the WFD considers as the most important to assess the status of surface water bodies, are a new element in EU water management. The diversity and abundance of the fauna and flora in a water body have to be examined and monitored. The EU Member States are obliged to define water body types and type specific biological reference conditions.

Economic aspects are also relevant in the WFD. The Directive requires the recovery of costs for water services including environmental and resource costs. EU Member States shall ensure by 2010 that water-pricing policies provide adequate incentives for efficient water use and that industry, households and agriculture contribute adequately to the costs of water services.

H. Vogtmann and N. Dobretsov (eds.), Transboundary Water Resources: Strategies for Regional Security and Ecological Stability, 121–127.

The Directive has to be implemented in different steps. Until December 2004 an analysis of the characteristics of the river basin district, a review of the impact of human activity on the status of water bodies and an economic analysis of water use are required. On the basis of these analyses and reviews the EU Member States have to develop monitoring programmes until December 2006 at the latest and later on they have to decide on the necessary measures in order to achieve good status of all water bodies. The main instruments of the Directive are national programmes of measures and national or international river basin management plans, which have to be established until December 2009 for each river basin district. Public participation and consultation play an important role in the production of programmes and plans. The measures have to be implemented until 2012, after that the instruments are reviewed and if necessary updated. The European Commission on the basis of reports by the EU Member States controls the implementation process.

The legal and technical implementation of the WFD is under way. A lot of questions have still to be answered. In order to facilitate and to harmonize the implementation process the European Commission and the Member States have developed several guidance papers with regard to important WFD requirements (analysis of pressures and impacts, economic analysis, public participation etc.).

INTRODUCTION

The European Water Framework Directive establishes a strategic framework, which aims at an integrated and sustainable management of surface and groundwater bodies in the European Union (EU). Programmes of measures and river basin management plans have to be produced in each river basin district in order to achieve good ecological, chemical and quantitative status. Coordination and cooperation within the EU Member States and among them are required. Therefore the Directive might provide model elements or approaches also for the transboundary water management in Central Asia.

1. THE WATER FRAMEWORK DIRECTIVE – GENESIS OF A NEW APPROACH IN EU WATER MANAGEMENT

The Water Framework Directive (WFD) came into force on 22 December 2000[1] after four years of intensive discussion in the European Council and the European Parliament. Until then EU water law was a piecemeal of a lot of different directives dealing with special issues of water use and management like waste water treatment, drinking water or water pollution by dangerous substances. The States cooperated already on international or bilateral level, e.g. in international river basin commissions, but in a more general and less binding way. Against this background the EU Member States asked for a more coherent and integrated water management strategy based on the view of waters as transboundary ecosystems.

The WFD introduces new approaches in EU water management. River basin districts are the main management units in the future. According to the WFD definition "river basin district means the area of land and sea, made up of one or more neighbouring river basins together with their associated groundwaters and coastal waters". Germany for example participates in the international river basin districts of the Rhine or the Danube. The management of river basin districts requires more intensive cooperation and coordination within and among EU Member States and Non Member States in order to meet the provisions of the different WFD implementation steps. This includes new cooperation and coordination structures and procedures. For the first time ecological criteria play an important role with regard to water quality. Until then mainly the good chemical status of a water body was relevant in EU water law. Additionally the WFD requires the broad information, consultation and active involvement of the public in the implementation process, which is also a new element in water management in a lot of EU Member States.

2. IMPLEMENTATION OF THE WFD

2.1 Implementation Time Table

The WFD has to be implemented step by step.

Deadline	Task
December 2003	Transposition of the WFD into national law
December 2004	Analysis of the current status of the water bodies in a river basin district: Characteristics of the river basin district Review of the impact of human activity on the status of water bodies Economic analysis
December 2006	Establishment of monitoring programmes
After December 2006	Analysis of the deficits: Which water body does not achieve the required good status? Which measures have to be taken to achieve good status?
December 2009	Development of programmes of measures and management plans for each river basin
December 2015	Good status of water bodies to be achieved (This deadline may be extended until 2027)

At the moment the EU Member States analyse the status of the water bodies. In this implementation phase all existing data and information on the status have to be reviewed, e.g. the human activities which have an impact on water bodies, like water abstractions, water flow regulations, point source pollution from industrial or agricultural installations and activities.

2.2 Environmental objective of the WFD: Good status

The main environmental objective of the WFD is to achieve the good status of all water bodies in European river basin districts until December 2015.

Good status means good ecological and chemical status of surface water bodies (rivers, lakes, transitional and coastal waters). With regard to the ecological status diversity and abundance of typical water fauna and flora are essential criteria for the classification of the status of a water body. EU Member States have to define surface water body types on the basis of inter alia geographical and geological criteria. Type-specific reference conditions have to be established for each type with regard to biological (macrophytes and phytobenthos, fish fauna etc.), hydromorphological (hydrological regime, river continuity etc.) and chemical (special pollutants) elements.

Groundwater bodies shall achieve good chemical and quantitative status. Good quantitative status means that the level of groundwater in the groundwater body is such that the available groundwater resource is not exceeded by the long-term annual average rate of abstraction.

The deadline of 2015 may be extended until 2027 if e.g. good status can not be achieved earlier for reasons of technical feasibility. The Member States have to meet rather stringent conditions to make use of this prolongation possibility.

The WFD also allows limited derogations concerning the environmental objectives. Member States may designate a water body as artificial or heavily modified if there are necessary changes to the hydromorphological characteristics of a surface water body to enable e.g. navigation. In this case the good ecological potential has to be achieved instead of the good ecological status, i.e. that only those measures should be taken to improve water ecology, which are compatible with navigation. Additionally the WFD derogations also cover setting lower objectives, allowing for new development, and temporary deterioration in quality. The WFD also foresees that Member States may aim to achieve less stringent objectives if a water body is so affected by human activity that it would be infeasible or disproportionately expensive to achieve good status.

2.3 Objectives, measures and instruments to achieve good status

The WFD requires that further deterioration of the current status of water bodies shall be prevented. That means e.g. that the competent authority cannot authorize the abstraction of groundwater if that has negative effects on the quantitative status of a groundwater body. Water bodies shall be protected and restored.

Pollution of water bodies has to be prevented or limited. The Member States have already to achieve compliance with emission limit values or environmental quality standards for a number of pollutants in EU environment directives. Additionally the WFD foresees the establishment of limit values and quality standards for so called priority substances, i.e. pollutants that are toxic, persistent and liable to bio-accumulate. There is a list of ca. 30 substances at the moment, like lead or mercury and some pesticides.

The main instruments of the water framework directive to achieve good status of water bodies are the programmes of measures and the river basin management plans. Both have to be established for the first time until December 2009 and after that in a 6-year rhythm.

The programme of measures has to be established on national level by the Member States for each river basin district. There will be e.g. a German programme of measures for the German part of the Danube river basin. These programmes have to be coordinated with other Member States in the same river basin district. They consist of minimum basic measures and supplementary measures. Basic measures include those required to implement all relevant EU legislation for the protection of water, for example, the Urban Waste Water Treatment Directive. Other basic measures include controls over water uses, e.g. prior authorization of the abstraction of groundwater or surface water, as well as controls on point source discharges. Besides the basic measures the Member States have a lot of leeway in the choice of suitable measures to achieve good status. The WFD also provides for Member States to adopt supplementary measures like environmental agreements or recreation of wetlands if necessary to achieve the environmental objectives.

The river basin management plan is the integrated management concept of a river basin district. In this plan the results of the different implementation steps will be presented in a summarized form: Analysis and evaluation of the status quo, definition of environmental objectives for the water bodies including derogations, monitoring programmes, summary of the programmes of measures, presentation of the public participation measures and their results. The management plan is either a national or in the case of a transboundary river basin an international plan. In the latter case this plan has to be coordinated among the Member States. The river basin management plan and its development is the basis for a three phased public consultation procedure and it is the instrument for the European Commission to control the implementation process.

2.4 Economic aspects in the WFD

According to the WFD an economic analysis have to be undertaken until December 2004. This analysis consists inter alia of calculations taking into account long-term forecasts of supply and demand for water. The

Member States have to estimate volume, prices and costs of water services and the relevant investments in the future. Water services are, for example, the abstraction, storage, treatment and distribution of surface water or groundwater. Judgements about the most cost-effective combination of measures to achieve good water status shall also be delivered in this analysis based on the estimated potential costs.

The Member States shall as far as possible achieve recovery of costs for water services taking into account the polluter pays principle and environmental and resource costs. By 2010 Member States shall ensure that the national water pricing policy provides incentives for users to use water efficiently. Industry, households and agriculture, i.e. the three main sectors of water users, shall contribute adequately to the recovery of costs by then.

2.5 Harmonised implementation of the WFD in the EU

The WFD is a rather complex directive and leaves a lot of questions unanswered. A harmonised implementation of the WFD within the EU is important, because of the transboundary river basin districts. The basis of analysis and evaluation must be similar in the Member States in order to achieve compatible and comparable results. Therefore the Common Implementation Strategy of the European Commission and the Member States has been established. Several committees have discussed and are still discussing the most relevant WFD requirements in order to develop guidance papers as a basis for the EU wide harmonised implementation. 15 Guidance documents e.g. on the economic analysis, on the identification of water bodies or on the analysis of pressures and impacts on water bodies have already been agreed on[2].

3. ORGANISATIONAL ISSUES

The WFD requires national and international cooperation and coordination to ensure an integrated management of all water bodies in a river basin district. Therefore coordination bodies and procedures have to be established. The programmes of measures and the river basin management plans need a bottom-up and a top-down coordination: The analyses and evaluation of the water body status have to be undertaken by the relevant local or regional authorities and then collected and summarized on national and/or international level. The provisions, the framework and the timetable have to be agreed on national and international level.

New bodies and committees have been created on national and on international level often using the already existing international river basin commissions (like for the Danube, the Rhine, the Oder) as basis. The restructuring of existing commissions is considered, which will require the adaptation of international treaties or agreements. The coordination structures should also take into account conflict management procedures to deal with situations when states within a river basin district can not agree e.g. on the measures necessary to achieve good status.

REFERENCES

1. Official Journal of the European Communities L 327, 22.12.2000, p. 1, see also WFD Homepage of the European's Commission Environment Directorate http://europa.eu.int/comm/environment/water/water-framework/index_en.html

2 EU CIRCA Homepage on WFD Implementation http://forum.europa.eu.int/Public/irc/env/wfd/information (see "library" for the guidance documents)

THE ROLE OF ECONOMICS TO PROMOTE A SUSTAINABLE USE AND MANAGEMENT OF TRANSBOUNDARY WATER RESOURCES

HELLEGERS, P.
Agricultural Economics Research Institute, The Hague, Netherlands

Key words: economic efficiency, social equity, sustainability, demand management, policy instruments

ABSTRACT

The last decade there has been a growing interest in the idea of "treating water as an economic good". There is, however, a lot of confusion about the meaning of "treating water as an economic good". It relates to making the right choices about optimal use and optimal allocation of water among users on the basis of socio-economic trade-off analysis. Insight into the value of water in alternative uses is important for making the right choices. This is different from water pricing, which is an economic instrument that can be used to achieve policy objectives, such as demand management or cost recovery. "Treating water as an economic good" does not automatically mean that water should be allocated by competitive market prices so that the available resource is fully allocated, and allocated to its highest-value uses as is often believed. There are other economic instruments that can be used as well such as tradable water rights, subsidies and block-rate tariffs. Regulatory instruments, such as rationing, and persuasive instruments, such as extension, can also be very cost-effective and suitable.

The suitability of instruments depends on the kind of water policy objectives, such as cost recovery, to ensure that supply and demand are brought into equilibrium or to reallocate water from less to more productive uses. Often more instruments are needed simultaneously and preconditions have to be fulfilled. Water is an economic good in the sense that it cannot fully satisfy demand for al its alternative uses simultaneously. Water is, however, also a social good whose availability to certain groups and for certain purposes will serve the greater benefit of society as a whole. Access to clean water is often seen as a basic right of all human beings. It is often considered as too vital to be left to the economic forces of profit-maximisation. Goals other than efficiency, like social equity and sustainability, are often guiding. This explains why the government often subsidises those uses of water that have a high value, but low ability to pay.

Economic instruments have a number of advantages. They increase the water use efficiency. Besides, they offer ongoing incentives to reduce

H. Vogtmann and N. Dobretsov (eds.), Transboundary Water Resources: Strategies for Regional Security and Ecological Stability, 129–140.

usage and to innovate. They are flexible –in the sense that they can be modified and adjusted easily-. Finally, they may generate revenues. In spite of these advantages, economic instruments are not widely applied in water resource management for a number of reasons. Firstly, there might be market imperfections, such as externalities. Secondly, there might be an uncertain relationship between charges and impact on water use. Water prices are often small compared to the value of water. A considerable increase in the price of water is needed to affect demand, which is often socially not desirable. Thirdly, they may not be widely applied because they are new or politically sensitive. Fourthly, transaction costs may be high relative to the size of the efficiency gains. Finally, preconditions for implementation are often not met, such as defined water rights or volumetric measurement. The role of economic instruments to promote a sustainable use and management of water is therefore currently limited. Economics mainly plays a role in the analytical part. Economics provides us with tools that may be useful in resolving competition among alternative uses.

The suitability of economic instruments for transboundary water management will be discussed on the basis of a case study. Egypt is mainly served by surface water from the river Nile and Lake Nasser. Egypt currently receives 68 bm3 of surface water per year and uses about 60 bm3, which exceeds the 55.5 bm3 of water 'agreed' with Sudan. When Sudan uses all its water entitlements, water supply will decline. Water availability in Egypt will be insufficient considering the high water demand for intensive cropping and plans of the government to expand the irrigated area by 40%. Besides there is a rapid increase in population and water demanding industry. The amount of water assigned to Egypt and Sudan is specified in quantity and time on the basis of rationing. Often this does not allocate water in an economically efficient way. Theoretically water entitlements –which can be considered as a kind of water use rights- can be re-allocated between countries through market mechanisms, i.e. introduce tradable water rights. The marketability of rights encourages users to reduce low value usage and sell surplus water. Such a reallocation is, however, politically sensitive and high transaction costs will be involved

1. INTRODUCTION

The last decade there has been a growing interest in the idea of treating water as an economic good, as highlighted at several conferences. The Second World Water Forum (The Hague, March 2000) stressed that decisions on water allocation among competing uses require a better analysis of the value of water (SWWF, 2000). The International Conference on Water and the Environment (Dublin, January 1992) emphasised that failure to recognise the value of water has led to environmentally damaging uses of the resource (ICWE, 1992). The United Nations Conference on Environment and Development (Rio de Janeiro, June 1992) underlined that the role of water as a social and economic good should be reflected in demand management mechanisms (UNCED, 1992).

International bodies (OECD, 1989; World Bank, 1993; FAO, 1994) also recognise the need to study the optimal allocation of water. Especially the allocation of shared water resources among countries, given the increase in competition for available freshwater resources. The availability of water, and access to its utilization, is crucial to the economic well being not only of individuals, but also of entire countries.

Special attention will be given in this paper to irrigated agriculture, as it is the largest consumer of water and the use of irrigation water can cause serious negative effects on the environment. About 72% of the world water abstractions are used for the production of food and fibers. Water is an important production factor: irrigated agriculture provides about 40% of the world's food supply, but occupies only 17% of the cultivated area (OECD, 2002).

An irrigation disaster is for instance the Aral Sea. In the 1950s Soviet planners diverted large parts of the two rivers that feed the Aral Sea –the Amu Darya and the Syr Darya- to irrigate cotton crops in the near-desert terrain of central Asia. The Aral Sea soon started to dry up. Since 1960, it has shrunk by three-quarters in volume. Almost all the fish has died out. Moreover, rapidly rising salinity has killed many crops. All this for a few million tons of heavily subsidized cotton produced. The Aral Sea may never recover though its shrinkage seems to have slowed recently (The Economist, 2003).

The main aim of this paper is to discuss the role of economics to promote a sustainable use and management of transboundary water resources. Economics provides us with two contributions that may be useful in this respect: economic analysis and economic instruments. Economic trade-off analysis, like cost-benefit analysis and cost-effectiveness analysis, can be used to help policy makers in making the right decisions on the allocation of water, whereas economic instruments influence human behaviour indirectly by providing incentives to use water more efficiently.

The structure of this paper is as follows. In Section 2 the meaning of treating 'water as an economic good' will be explained. Review criteria for socio-economic trade-off analysis to determine optimum situations will be discussed. In Section 3 the difference between the price, value and costs of water will be explained. Water pricing is, however, not the only policy instrument that can be used to achieve optimum situations. In Section 4 alternative policy instruments to achieve optimum situations will be discussed as well as the possible role of economic instruments in transboundary water management. Finally, in Section 5 concluding remarks are drawn.

2. MEANING OF TREATING WATER IN IRRIGATED AGRICULTURE AS AN ECONOMIC GOOD

The meaning of treating 'water in irrigated agriculture as an economic good' relates to making the right choices about optimal use and optimal allocation of water among potential users on the basis of socio-economic trade-off analysis (independent of the ability to pay). Insight into the value of water in alternative uses and in different countries, which share water resources, is important for making the right choices about optimal use and optimal allocation of water as a scarce resource. Treating 'water as an economic good' is about making the right choices, and not about setting the appropriate price for water, as it is often believed (Savenije, 2000). It does not mean that water should be allocated by competitive market prices (Perry et al., 1997).

The economic efficiency criterion is only one of the basic principles for making socio-economic trade-off analysis. The criterion of efficiency is attractive to economists because it carries little ethical content. There is, however, no reason to believe that resource allocations are socially desirable just because they are efficient. Optimality can only be assessed in terms of social welfare, which is only meaningful if there are agreed ethical principles. An optimal arrangement is efficient, but an efficient arrangement is not necessarily an optimal one since other criteria, such as social equity and ecological sustainability, also may play a role. The most important review criteria for socio-economic trade-off analysis are discussed below.

-*Economic efficiency* is achieved when the net marginal productivity of water is equal for all countries. Water transactions are interesting if differences exist between the net marginal productivity.

-*Social equity* shows whether the costs and benefits associated with changes in the allocation of water are equitably distributed among countries with shared water resources.

-*Environmental sustainability* is often a prerequisite for optimal water use in the long run. Depletion of groundwater aquifers over a long period of time is for instance not sustainable.

Many past failures in water resources management are attributable to the fact that water has been viewed as a *free good*. This may lead to water being allocated to low-value uses and provides no incentives to treat water as a limited asset. Water is an *economic good*, since it is a scarce resource that can be allocated in alternative ways. At least to some minimal level of availability, water is also a *social good* whose availability to certain groups and for certain purposes will serve the greater benefit of society as a whole. Access to clean water is often seen as a basic right of all human beings (ICWE, 1992), because it is often considered as too vital to humans to be left to the economic forces of profit-maximization (Gibbons, 1986). Besides, many societies believe that water has cultural and religious values

(FAO, 1994). Goals other than economic efficiency are often guiding principles. This explains why the government sometimes subsidizes those uses of water that have a high value, but low ability to pay (Hartwick and Olewiler, 1998). Social concerns may require subsidies, but need to be transparent. It is therefore a challenge to identify the right balance between water treated as an economic good and water treated as a social good.

Public intervention in the water sector is often inevitable due to market failure, which is the divergence between the market outcome (without intervention) and the economically efficient solution. Market failures in water resource management fall into three major categories:

(1) *Externalities* are unintended side effects of one party's action on another party (i.e. country) that are ignored in decisions made by the party causing it. Countries down-stream may for instance be affected by water diversions or water pollution by countries up-streams. The consequences of not internalising such externalities in the water price are shown in Section 3.
(2) In many irrigation projects, the physical and management infrastructure required to price water at its marginal costs is absent and *transaction costs* of such infrastructure are high.
(3) The greatest problem is that *property rights* are often not well defined. Coase (1960) stresses the importance of the existence of well-defined property rights. He reduces market perfection conditions to two issues and shows that market allocation will be efficient given well-defined water rights and low transaction costs, regardless of the allocation of rights. Well-defined water rights are, however, hard to establish, since water is not a homogeneous product. It is important to note in this respect that a well-managed irrigation sector needs a balance between private and public involvement.

3. WATER PRICING

Treating "water as an economic good", is completely different from water pricing. Water pricing can be used as a policy instrument for demand management and cost recovery. It can be used to encourage countries to make better use of their water resources and to ensure that the service costs are completely borne by the users, as will be explained below in more detail.

Water scarcity is a relative concept. In most countries the imbalance between water demand and supply can be bridged through management reforms. Challenges posed by growing scarcity can be addressed through two strategies: *supply management*, which involves developing new supplies, and *demand management*, which promotes policies that make better use of existing supplies (Winpenny, 1994). When water becomes increasingly scarce, continuing the traditional policy of extending supply is no longer a feasible option. There is a need for instruments that can contribute to management by limiting the demand for water.

The full cost of providing water includes the full economic cost and the cost of environmental externalities associated with public health and ecosystem maintenance (see Figure 1). The full economic cost consists of the full supply costs, the opportunity costs from alternative water uses and the economic externalities arising from changes in economic activities. The full supply costs consist of operating and maintenance expenditures and capital charges.

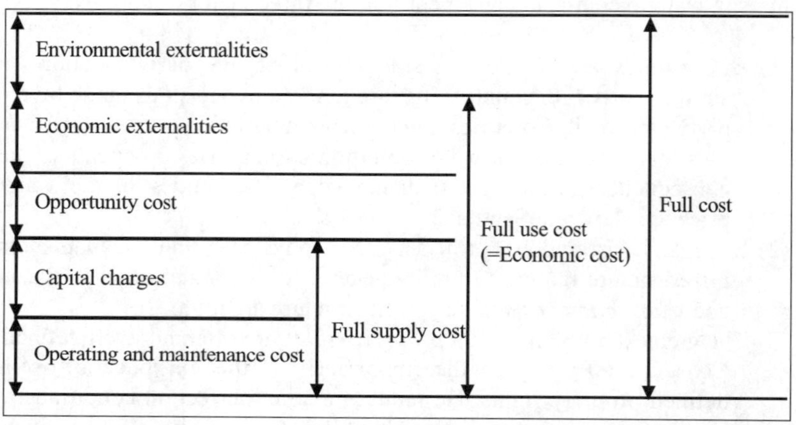

Figure 1. Concept to determine the full cost of the provision of water (after: GWP, 2000).

There are large differences among countries in the way the price of water covers costs. In some countries, the price does not even cover operation and maintenance costs. In other countries capital costs are included as well, and sometimes an effort is made to include opportunity cost and cost of externalities in the price of water (Dosi and Easter, 2000).

The impact of not considering the negative externalities of water use in the price of water when making extraction decisions is shown in Figure 2. If there are negative externalities, the marginal social costs of water use will be higher than the marginal private costs. Farmers will use water as long as the private benefits of an additional unit exceed private costs. This means that farmers will demand the quantity of water E^p at a price v^p. From a social point of view, farmers extract too much in the presence of negative externalities and water is under-priced. External costs that have to be borne by society in the private optimum E^p are shown by the triangle ABC. If externalities are internalised in the price of water, farmers will face a price v^* and extraction will be reduced to the socially optimal extraction level E^*, where marginal benefits equal marginal social costs. Figure 2 shows that the decrease in costs (the area DBE^pE^*) exceeds the decrease in benefits (the area DCE^pE^*) and there are social welfare gains (shown by the triangle DBC). Figure 2 shows that at the price v^p, a tax level T^v can reduce extraction to the socially optimal level E^*. The shaded rectangle shows the tax revenue.

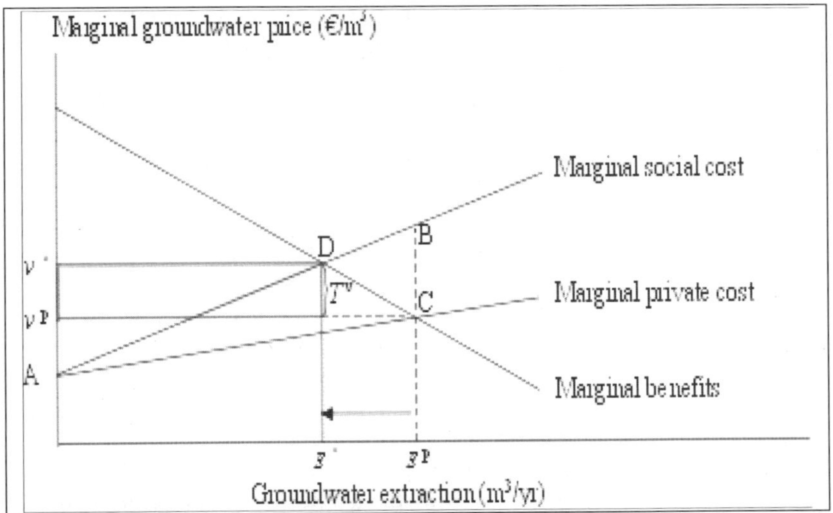

Figure 2. Social welfare gains of the internalisation of negative externalities in water price.

The *value* and *cost* of water should not be confused with the *price* of water, which is the amount of money actually paid for it (Rogers et al., 1998). The full value is the sum of use (economic) values and non-use (intrinsic) values. The price does often not reflect the value and cost of water. Often there is a 'rent', which is the difference between the actual price and the value of water, i.e. the maximum amount the user would be willing to pay. When the price of water reflects its true costs, the resource will be put to its most valuable uses (Rogers et al., 2002). In practice, the actual price paid is often lower than the value, because social and political goals often override economic criteria.

4. ALTERNATIVE POLICY INSTRUMENTS TO ACHIEVE OPTIMAL SITUATIONS

Treating water as an economic good does not mean that economic instruments, such as higher prices, should automatically be used in practice to achieve optimal situations, as is often believed (Savenije, 2000). Other instruments can also be very cost-effective and suitable. Various policy instruments for water resource management can be used by the government to achieve optimum resource management. They are classified here in two categories:

1) *The institutional and legal environment,* in which water is supplied and used, can be changed. Institutions set the 'rules of the game' and help to define the rights, privileges and responsibilities that guide human activities. The water rights

system is a water institution since it includes provisions that determine access to water (Bromley, 1991).

2) *Environmental policy instruments for water management* are usually divided into market-based and non-market-based devices (see Winpenny, 1994; Rosegrant, 1997). Since non-market-based devices take a variety of forms, a distinction is made between regulatory and persuasive instruments. The following division into three classes – based on the way in which a government can influence an actor's behaviour – is often made.

- *Economic instruments* are market-based. They influence indirectly the behaviour by providing incentives to use water efficiently. Three types can be distinguished:
- Charges or taxes: a straightforward way to put prices on the use of the environment.
- Tradable rights or marketable permits: environmental quotas or ceilings on extraction that, once initially allocated by authorities, can be traded subject to prescribed rules.
- Subsidies (usually in the form of grants): in general, these are incompatible with the Polluter-Pays Principle and can only be justified as a transitional measure.
- • *Regulatory instruments* (also known as command-and-control instruments) can directly restrict, forbid and/or prescribe action. The main feature is that there is no legally free choice for actors, who face penalties. Examples are permits, standards and rationing.
- • *Persuasive instruments* work even more indirectly than economic instruments, because actors are supposed to take action of their own accord. There are no sanctions. Examples are education, extension, information, negotiation and voluntary agreements.

Policy instruments for demand management that have been widely studied under different socio-economic and physical environments are water pricing, tradable rights and quantity-based control. The advantages of tradable rights and quantity-based controls will be discussed here. See Bhatia et al. (1994), FAO (1994) and Winpenny (1994) for more discussion.

Tradable water rights are appealing since they combine an advantage of the quantity-based system with the cost advantage of a tax system (Pindyck and Rubinfeld, 1989). The agency that administers the system determines the total number of rights and therefore the total amount of extraction, just as quantity-based systems would do. But the marketability of the rights allows extraction reduction to be achieved at minimum cost, just as a system of taxes would do. The objective will still be reached even if new parties enter the market, since the total number of rights is limited. Also, it offers nature organisations the opportunity to buy rights in order to reduce extraction (Pearce and Turner, 1990). Since all requirements for well-functioning water markets (such as water scarcity, well-defined and transferable rights and limited transaction costs) are often not met and as a result of potential third-party effects of transaction that may arise, water markets have failed to develop in many areas.

Quantity-based control mechanisms, like rationing by means of extraction quotas, constrain extraction to the socially optimal level, by imposing substantial monetary penalties for exceeding the limits. If the price elasticity of water demand is equal or close to zero, quotas will be more effective in constraining extraction than water pricing. If the price elasticity of water demand is significantly different from zero and negative, quotas and pricing lead to the same extraction reduction when the demand is constant. However, if increases in demand for water are expected, then a quota is preferred as it constrains extraction at the same level, while extraction will change under water pricing.

Economic instruments have a number of advantages (OECD, 1991). Firstly, they are static efficient, which means that extractions occur in countries where it is most efficient to do so. Secondly, they offer an ongoing incentive to reduce extractions below the level that environmental policy prescribes. The property of economic instruments to offer more incentives to innovate than direct regulation is called dynamic efficiency (Dijkstra, 1998). Thirdly, they increase flexibility. It is generally easier to adjust a tax than to change legislation. Finally, they may generate revenues for the government.

In spite of these advantages, economic instruments are not widely applied in environmental policy for a number of reasons (cf. Dijkstra, 1998). Firstly, there might be market imperfections, such as externalities. Secondly, there might be an uncertain relationship between charges and impact on water use. Water prices are often small compared to the value of water. A considerable increase in the price of water is needed to affect demand, which is often socially not desirable. Thirdly, they may not be widely applied because they are new or politically sensitive. Fourthly, transaction costs may be high relative to the size of the efficiency gains. Finally, preconditions for implementation are often not met, such as defined water rights or volumetric measurement.

General guidelines for the use of policy instruments are difficult to establish, since the suitability of policy instruments depends on the characteristics of the situation and on the kind of water policy objectives, such as cost recovery, to ensure that supply and demand are brought into equilibrium or to reallocate water from less to more productive uses. Criteria for policy review are economic efficiency, effectiveness (to what extent do policies achieve their objectives), administrative feasibility (how easy is it to implement, monitor and enforce policies), social equity and acceptability. It is important to note that policy instruments can be combined in such a way that they reinforce each other. Although these policy mixes are complex and difficult to develop, the most successful experiences indicated that they are viable and perhaps the only way to achieve multiple-objective reforms. This brings OECD (2002) to conclude that a mixture of instruments can be fruitful and needs to be exploited.

The suitability of economic instruments for transboundary water management is discussed here on the basis of the Nile basin. Egypt is mainly served by surface water from the river Nile and Lake Nasser. Egypt currently receives 68 bm^3 of surface water per year, which exceeds the 55.5 bm^3 of water 'agreed' with Sudan. When Sudan uses all its water entitlements, water supply will decline. Water availability in Egypt will be insufficient considering the high water demand for intensive cropping and plans of the government to expand the irrigated area by 40%. Besides there is a rapid increase in population and water demanding industry. The amount of water assigned to Egypt and Sudan is specified in quantity and time on the basis of rationing, which does not allocate water in an economically efficient way. Theoretically water entitlements –which can be considered as a kind of water use rights- can be re-allocated between countries through market mechanisms, i.e. tradable water rights.

The marketability of rights encourages users to reduce low value usage and sell surplus water. Such a reallocation is, however, politically sensitive and high transaction costs will be involved.

The potential role of market forces to allocate water among countries seems high when there are large differences in the net marginal productivity of water among countries. It is, however, not likely that the government will allow market forces to fully divert water away from usage with a low net marginal productivity or from usage that is hard to express in monetary term – not least because many of the lowest (economic) value uses have the highest socio-political value, like wetlands. Generally the basic allocation of water among countries is a political decision, because allocation has so many implications that the potential uncertainties of a 'free market' solution are politically unacceptable.

CONCLUDING REMARKS

The main aim of this paper was to clarify the role of economics to promote a sustainable use and management of transboundary water resources. It became clear that economics is useful in tracing through the implications of various options for allocating scarce water resources among countries that share water resources. It is important to note in this respect that the economic efficient allocation is not necessary the social optimum one, since other criteria such as social equity and environmental sustainability may play a role. Goals other than economic efficiency are often guiding principles.

Generally the basic allocation of water among countries is a political decision, because allocation has so many implications that the potential uncertainties of a 'free market' solution are politically unacceptable. The role of economic instruments to promote a sustainable use and management of transboundary water resources is therefore currently limited. Public intervention in the water sector is often inevitable, especially when water use causes negative externalities on the environment, like in the case of the Aral Sea. Economics can play a role in internalising such externalities of water use in the price of water.

REFERENCES

1. Bhatia, R., R. Cestti and J. Winpenny, 1994, Policies for water conservation and reallocation: good practice cases in improving efficiency and equity. World Bank, Technical Paper, Washington.
2. Bromley, D., 1991, Environment and economy: property rights and public policy. Basil Blackwell Ltd, Oxford.
3. Coase, R., 1960, The problem of social cost. Journal of Law and Economics, 3:1-44.
4. Dijkstra, B.R., 1998, The political economy of instrument choice in environmental policy. PhD thesis, University of Groningen, Groningen.
5. Dosi, C. and K.W. Easter, 2000, Water scarcity: economic approaches to improving management. International Food and Agricultural Policy, WP00-2, St.Paul.
6. FAO, 1994, Reforming water resources policy: a guide to methods, processes and practices. Food and Agriculture Organisation, Irrigation and Drainage Paper No.52, Rome.
7. Gibbons, D.C., 1986, The economic value of water: resources for the future. Washington.
8. GWP, 2000, Integrated water resources management. Technical Advisory Committee, Global Water Partnership, Paper No.4, Stockholm.
9. Hartwick, J.M. and N.D. Olewiler, 1998, The economics of natural resource use. Addison-Wesley, Reading.
10. ICWE, 1992, The Dublin statement and report of the International Conference on Water and the Environment. Development issues for the 21st century, 26-31 January, 1992, Dublin.
11. OECD, 1989, Water Resource Management: integrated policies. OECD, Paris.
12. OECD, 1991, Environmental policy: how to apply economic instruments. OECD, Paris.
13. OECD, 2002, Transition to full-cost pricing of irrigation water for agriculture in OECD countries. OECD, Paris.
14. Pearce, D. and R. Turner, 1990, Economics of natural resources and the environment. Harvester Wheatsheaf, Hertfordshire.
15. Perry, C.J., M.Rock and D. Seckler, 1997, Water as an Economic Good: A solution, or a Problem? Research Report 14, International Irrigation Management Institute, Sri Lanka.
16. Pindyck, R.S. and D.L. Rubinfeld, 1989, Microeconomics. Mac Millan, New York.
17. Rogers, P., R. Bhatia and A. Huber, 1998, Water as a social and economic good: How to put the principle into practice. Global Water Partnership / Swedish International Development Co-operation Agency, Stockholm, Sweden.
18. Rogers, P., R. de Silva and R. Bhatia, 2002, Water is an economic good: how to use prices to promote equity, efficiency and sustainability. Water Policy, 4:1-17.
19. Rosegrant, M.W., 1997, Water resources in the twenty-first century: challenges and implications for action. IFPRI, Discussion Paper No.20, Washington.

20. Savenije, H., 2000, The socio-economic value of water: knowledge gaps and research questions in integrated water resources management. Paper presented at the UNESCO-WOTRO Conference on Water for Society, 8-10 November, 2000, Delft.

21. SWWF, 2000, Forum report on water and economics of the Second World Water Forum, The Hague.

22. The Economist, 2003, Priceless. Irrigate and die, under-pricing and misallocating water carries a high price, July 19th, 2003.

23. UNCED, 1992, United Nations Conference on Environment and Development, Final communication. Rio de Janeiro.

24. Winpenny, J., 1994, Managing Water as an Economic Resource. Routledge, London and New York.

25. World Bank, 1993, Water resources management: a World Bank policy paper. World Bank, Washington.

SATELLITE ALTIMETRY FOR MONITORING LAKE LEVEL CHANGES

CRETAUX, J.-F., A. KOURAEV, M. BERGE-NGUYEN, A. CAZENAVE & F. PAPA
CNES / LEGOS, Toulouse, France

ABSTRACT

Accurate and continuous monitoring of lakes and inland seas is possible since 1991 thanks to the recent missions of satellite altimetry (Topex-Poseidon, ERS-1, ERS-2, Jason-1 and Envisat). Global processing of the data of these satellites could provide temporal and spatial times series of lakes water level from 1991 to 2003 on the whole Earth with a decimeter precision. The response of water level to regional hydrology is particularly marked for lakes and inland seas of semi-arid regions. Altimetry data can provide an invaluable source of information in hydrology sciences, but in-situ data (rivers runoff, temperature, precipitation etc.) are still strongly needed to study the evolution of water mass balance of each lake. Moreover, sea level variations that result from variation of hydrological parameters such as river discharge, precipitation and evaporation, are very sensitive indicators of regional climate variations. Recent results obtained on Aral Sea and Issykkul Lake are presented here. Inter - annual changes of water level have been obtained over these lakes that must be interpreted in term of hydrological water balance. Since 1960 the Aral sea has been drying and since 1989 it is divided into two lakes that follow different evolution, the Big Aral in the south which continuously dried up the last 10 years, while the so-called Small Aral in the north presented large inter-annual fluctuations related to constructions and destructions of a dam in the Berg's strait retaining the water from the Syr Darya. For Issykkul, a slow decrease of the level has been observed over the last hundreds years (4 cm / year), followed by an abrupt and bigger increase of the level of around 10 cm/yr since 1998. The impact on local populations and infra-structures of these fluctuations are dramatic in the case of Aral, much less in the case of Issykkul, but comparative study of both water bodies may help in the future to understand the respective consequences of human-induced activities from the natural changes. It is also the task of a new project recently submitted and accepted by the NATO with scientists from Uzbekistan and Kyrgyz Republic.

INTRODUCTION

In August 1992, the TOPEX/Poseidon (T/P) satellite has been launched on a 66° inclined orbit at 1336 km altitude with objective of measuring ocean height at a very high accuracy of few centimeters. The

H. Vogtmann and N. Dobretsov (eds.), Transboundary Water Resources: Strategies for Regional Security and Ecological Stability, 141–146.

orbit repeat period is 10 days which corresponds to an inter track spacing of 250 km at equator. Altimetry, although designed to study open ocean, has been immediately used over continental lakes (for exhaustive details on application of altimetry to lake studies, see Birkett 1995). The satellite carries a dual frequency radar altimeter operating in C and Ku bands (5.3 and 13.6 GHz respectively for T/P), which transmits a short pulse in the nadir direction, which is then reflected by the sea surface. The measurement of the time for pulse to be reflected back to the altimeter corresponds to the distance between satellite and sea surface (Fu & Cazenave 2001). Little correction must be applied to these measurements, due to atmospheric refraction, various electromagnetic bias, tides, and then, instantaneous sea surface height (SSH) is simply obtained from the difference between the radial component of the orbit and the corrected altimeter measurement. For previous altimeter missions, uncertainty in the radial orbit component was the largest source of error affecting SSH. For T/P, the orbit precision is currently 2-3 cm rms (Nouel et al., 1994) and thus it allows a high precision altimetry. This technique can be used on all water surfaces, ocean as well as lakes, inland seas, flooding plains, or rivers. It is just requested that the projection of the orbit of the satellite cross the surface of the water body and that this intersection is large enough to generate the return radar signal. The spatial resolution along the ground track is of about 6 km for 1-second measurement. This technique has been already applied to some large lakes since the last 5-6 years (Ponchaut & Cazenave 1998). Results have been obtained on Caspian Sea (Cazenave et al., 1997), on Mediterranean and Black Sea (Cazenave et al., 2002), African lakes (Mercier et al., 2002), and recently on Aral Sea (Aladin et al., 2003). In this article, we present results obtained recently on Big and Small Aral Sea and Issykkul lake.

ARAL SEA

The Aral sea, which was one of the world's largest inland water-bodies, with a surface of 66000 km^2 and a volume reaching more than 1000 km^3, started to shrink at the beginning of the 1960's when anthropogenic demands for agricultural needs led authorities to increase irrigation intake of the 2 inflow water contributors: Amu Darya and Syr Darya rivers. In 1960, the level of the sea was 53 m above the Baltic Sea level (taken as usual reference of zero sea level). Amu Darya and Syr Darya provided around 60 km^3 of fresh water runoff per year, which represents approximately half of the total runoff capacity flow of both rivers. The other half was lost by evaporation, underground infiltration lost and irrigation along the 3000 km length of the rivers. When withdrawal water for irrigation increased in 1960, the equilibrium of the water was broken, the level declined very rapidly and reached 39 meters in 1989 at the time of separation of Aral Sea into two parts: Small (northern part) and Big (Southern part) Aral Seas. During this period from 1960 to 1989, salinity of the sea has increased from 10 g/l to 28 g/l. Since 1989, both lakes have evolved differently. The Small Aral, which continued to receive water inflow from Syr Darya River, and suffer less from evaporation due to the small size of the lake, has fluctuated around the level 40 m, with highest level in 1998 when a dam has retained water from the river within the lake. The Big Aral received less water from Amu Darya, and due to its larger water surface, it has suffered from a high rate of evaporation during the last ten years. Using the T/P data, we have computed the variation of level and volume of both lakes (figures 1 and 2). In a first step we have averaged the level of the lakes deduced from all measurements made by T/P each ten days (Aladin et al., 2003). Then we used a dedicated Digital Bathymetry Map (DBM) of Aral Sea to compute time series of volume of Small and Big Aral with temporal resolution of 10 days.

In the same time we have computed their variations of volume taken into account of the in-situ hydrological data available through Internet and publications. For the runoff of Amu Darya and Syr Darya, we used data provided by Internet web site (http://water. freenet. uz) thanks to research made in the frame of an INTAS program (Ivan Stavitsky, personal communication). These corresponded to monthly average flow of Amu Darya (1993 to 2000) and Syr Darya (1993 to 1998) made at few dozens of kilometres upstream from the Big and Small Aral (In Kyzylgda and Kazalinsk, respectively). For evaporation and precipitation we have considered the values given in Small et al. (1999), that are based on outputs of a coupled regional climate-lake model (regCM2) for evaporation, and on Legate and Wilmott climatology data (Legates and Wilmott, 1990) for precipitation.

For Small Aral we have also taken into account that a dam has been built several times between 1993 and 1999 (Aladin et al., 2003). We have separated the period of treatment in 6 parts, 3 for the times when there was a dam in the Berg's strait, 3 for the times when the dam was destroyed. Comparing the time series of volume deduced from altimetry with variation of volume deduced from the hydrological budget (Precipitation plus Runoff minus Evaporation), we have estimated that the dam allow to retain 80 % of the Syr Darya river runoff passing the gauges of Kazalinsk. For the period of time without dam, we have estimated that only 20 % of river runoff reaches the sea. The rest is lost in evaporation in the delta and in the desert, as well as to underground infiltration. This computation allows quantifying the positive effect on water budget of the Small Aral .For the Big Aral, we have made the same computation. It first showed, that the volume of the lake measured by T/P measurements decreases slower than deduced from hydrological budget. To make both computations in agreement, we should consider an additional positive water inflow within Big Aral of around 5 km^3/year. This can be interpreted in different ways: errors of altimetry measurements, in the hydrological parameters we have used (Precipitation, Evaporation and Amu Darya Runoff), or lack of information about underground water flow. We have assessed the quality of altimetry measurement by comparing the surface of the lake (which can also be deduced from combination of altimetry and DBM) with satellite images of the Big Aral at different times between 1993 and 2003. This presented a very close agreement, which indicated that altimetry and DBM were accurate enough, and are not the source of disagreement within the hydrological budget.

Among the 2 other sources of errors (hydrological parameters, and possibility of non negligible underground inflow) it is up to now impossible to come to a definitive conclusion. This needs further analysis on each of these parameters. For evaporation, we need to take into account the high salinity of the lake (actually more than 80 g/l) which influences the rate of evaporation, as well as climate change since the time when the evaporation measurements were made (Small et al., 1999, 2001). For the runoff, there are possible errors, as we don't know how much water is lost between the gauge site used to compute Amu Darya runoff and the mouth of the delta. For possibility of underground water supply, this assumption needs hydrogeological investigations actually made, among other groups at the University of Neuchatel in Switzerland (Benduhn et al., 2003). This issue is far to be clear and solved.

LAKE ISSYKKUL

Lake Issykkul is an endorheic mountain lake located at 1608 m above the sea level in the northern Tian Shan, in the Kyrgyz Republic. It has an area of 6236 km^2, a length of 180 km, a width of 60 km, and a maximum depth of 668 m making it the fifth deepest lake in the world.

Rivers that flow into the lake are fed by melt-water from glaciers and snow, located above 3,300 meters altitude. The supply of river water into the lake result in the mean annual oscillation of lake level about 20 cm. The lake level progressively increases from February until beginning of September, decreasing afterwards progressively to the next February.

Historical sources (Holocene terraces, abrasion benches) point out that the lake level has strongly fluctuated. Thus, between the 11th BC and the 1^{st}AD centuries the lake level was 12 - 13 m higher than today, whereas between the 10th AD and the 12th AD centuries the lake level was lower than today. The 18th AD and the 19th AD centuries were also characterized by higher water levels than the present day (Mamatkanov et al., 2002).

During the last 75 years, the water level has progressively declined by about 4cm/year, and several factors have been invoked for explaining this phenomenon although the exact mechanism that trigger such oscillations remain unknown: the progressive increase of the intensive agricultural use of the lake shores together with the enlargement of the population inhabiting its shores (with consequent increase of the water consumption for domestic purposes) and the increase of the regional aridity conditions, are the main factors invoked for explaining the present day lake level drop (Mamatkhanov et al.,2002). From 1972 up to now, about 4 km^3 of glaciers have been thawing. For the period 1993 to 2003, as for Aral Sea, we have computed the variation of Issykkul with T/P altimetry data. As this lake is located in mountainous area, this decreases the accuracy of the altimetry measurements, mainly due to bigger errors made in the tropospheric correction applied to the range measurements, and the lost of many data due to the influence of surrounding topography of the lake region on the radar return signal. These limitations of altimetry for mountain lake have been described in Birkett et al., 1995, and possible source of improvements could be obtained through accurate atmospheric model used to compute the tropospheric correction, and retracking of the waveform of the altimetry radar echo that we didn't make for this analysis (Birkett et al., 1995).

However, since the last 4-5 years, measurements deduced from T/P altimetry mission have shown that the lake level of Issykkul increased with a rate of around 10 cm /year instead of dropping. In ten to twenty years, if lake Issykkul level continues to increase, damages on the shore infrastructure could be high. It can be the consequences of recent higher glacial melting. This could be connected to regional warming, which may also be registered in level variations of other neighbouring lakes like Karakul and Chatyrkul lakes.

PERSPECTIVES

In the scope of a NATO project, in coordination with Uzbek and Kyrgyz colleagues, we plan to investigate the inter-annual variability of Aral Sea and Issykkul Lake.

The main goal of the project is to precisely monitor variations of level and volume of both water bodies and to discriminate between global effect (consequences of climate changes on precipitation, evaporation, and glacier melting and regional or local effect (consequences of human activities: mainly withdrawal of water for agricultural consumption).

To outline natural and anthropogenic contribution to fluctuation of lake Issykkul, it is also proposed to study the level variations of other lakes, located in other parts of Holocene terraces that have different hydrological regime and different chemical composition. Lakes Karakul and Chatyrkul will be considered as proxies for this analysis. For example, Chatyrkul Lake is a mountainous lake, which is influenced only by climatic changes. Analysis of sediments, measurements of rivers runoff into these lakes and data of precipitation and lake temperatures will be used for the establishment of accurate current and historical water balance.

We plan to use various sources of information: in situ data of lake temperature, river runoff (from Amu Darya as well as rivers feeding Kyrgyz lakes), precipitation, evaporation, lake level measurement, etc.)and satellite altimetry data. This technique will allow to measure on a very short time scale the variations of average levels of lakes. Combination of different altimetry missions will also provide high temporal resolution of the mean sea level variations (10 days with T/P and Jason-1) and high spatial resolution of these variations (with ERS-1/2 and ENVISAT). It has been already applied to Aral Sea and Issykkul, with reliable results as shown above, that confirms and precisely measures the drying up of the Aral sea, and also provided continuous level decrease of Issykkul since 1998, followed by abrupt increase of its level. We also plan to use the

ERS1&2 data that provides a better density of measurements over the lakes and in case of Issykkul will help to not only consider average lake level variations, but also geographical distribution of water level fluctuations, that should give better estimates of currents and water redistribution within the whole lake.

REFERENCES

1. Aladin N, J-F Crétaux, I. S. Plotnikov, Kouraev A. V., A. O. Smurov, A. Cazenave, A. N. Egorov, F. Papa. "Modern hydro-biological state of the Small Aral Sea". *Environmetrics*, 2003, submitted.

2. Bendhun F and Renard P, A dynamic model of the Aral Sea water and Salt balance, *Journal of Marine Science*, special issue on Aral sea dynamic model, 2003

3. Birkett S., Contribution of TOPEX/POSEIDON to the global monitoring of climatically sensitive lakes, Journal of Geophysical Research, Vol 100, No C12, pp 25,179-25,204, 1995

4. Cazenave A., Bonnefond P., Dominh K., Caspian sea level from Topex / Poseidon altimetry: level now falling, Geophysical Research Letter, 24, 881-884, 1997

5. Cazenave A., Bonnefond P., Mercier F., Dominh K., toumazou V., Sea level variations in the mediterranean Sea and Black Sea from satellite altimetry and tide gauges, Global and Planetary Change, 34, 59-86, 2002

6. Fu L.L., and Cazenave A., Satellite altimetry and Earth Science, Academic Press, 2001

7. Legates R.D., and Wilmott C.J., Mean seasonal and spatial variability in gauge corrected, global precipitation, *Int J. Climatol.,* 10, 11127, 1990

8. Mamatkanov D.M., Bazhanova L.V., Dikikh A.N., Mandichev A.N., Matychenkov V.E., Romanovskiy V.V., Tuzova T.V., Shapar A.K. Water and Hydropower Resourses.// Mountains of Kyrgyzstan. Bishkek: Publishing house "Technologe". 2002. –pp.37-80.

9. Mercier F. A. Cazenave and C. Maheu, Interannual lake level fluctuations (1993-1999) in Africa from Topex-Poseidon: connections with ocean-atmosphere interactions over Indian Ocean, Global and Planetary Change, 32, 141-163, 2002.

10. Nouel F et al., Precise CNES orbits for TOPEX / POSEIDON : is 2 cm still a challenge, *J. Geophys res.,* 99, 24405-24420, 1994

11. Ponchaut F. & Cazenave A. (1998) 'Continental Lake Level Variations from Topex/Poseidon (1993-1996)', Comptes Rendus de l'Académie des Sciences, Sciences de la Terre et des Planètes, 326, 13-20.

12. Small E. E., Sloan L. C, Hostetler S., Giorgi F., Simulating the water balance of the Aral sea with a coupled regional climate-lake model, Journal of Geophysicall Research, Vol 104, No D6, pp 6583-6602, 1999

13. Small E. E., Giorgi F., Sloan L. C, Hostetler S., The effects of desiccation and Climate Change on the hydrology of the Aral Sea, Journal of Climate, Vol 14, pp 300-322, 2001

TREATY PRINCIPLES AND PATTERNS: SELECTED INTERNATIONAL WATER AGREEMENTS AS LESSONS FOR THE RESOLUTION OF THE SYR DARYA AND AMU DARYA WATER DISPUTE

DINAR, S.
Department of International Relations and Geography, Florida International University, 3000 NE 151st Street, 323A, Academic 1, North Miami, FL 33181, USA

ABSTRACT

After the collapse of the Soviet Union, Kyrgyzstan, Tajikistan, Uzbekistan, Turkmenistan, and Kazakhstan became independent states. The Amu Darya and Syr Darya Rivers, which were once under the auspices of one country, were now shared among five republics. Management of the two rivers and coordination of the water-sharing regime—once the responsibility of Moscow—became embroiled in the politics of newfound sovereignty. The water sharing regime and barter arrangements, which involved trading cheap fuel and electricity provided by downstream countries (Uzbekistan, Turkmenistan, and Kazakhstan) for water released by upstream countries (Kyrgyzstan and Tajikistan), was interrupted. Payments for reservoir upkeep were also halted. Additionally, Kyrgyzstan and Tajikistan have argued that they would like to develop their hydropower potential and receive monetary compensation for the inequitable water-sharing regime, favouring cotton production downstream. The five republics have achieved little progress in better sharing and managing their common waters and the current situation is far from satisfactory. In addition to discussing the water dispute and considering some of the interim progress made among the countries, this chapter will review lessons from other international water agreements that may provide a more adequate and long-term solution to the Central Asia water dispute. The chapter will specifically highlight two main principles (*compensation for facility use* and *compensation for downstream benefits*), which are expressed in several agreements that span conflicting water uses among upstream and downstream states. These principles can, in turn, be appropriately applied to the dispute over the Syr Darya and Amu Darya Rivers.

1. INTRODUCTION

The Central Asian dispute over the allocation of the Syr Darya and Amu Darya Rivers, and by extension the Aral Sea environmental crisis, are rooted in the last decades of history. While the two issues are related, the importance of water for national development has often meant that water allocation for agricultural production, namely cotton and rice, has

H. Vogtmann and N. Dobretsov (eds.), Transboundary Water Resources: Strategies for Regional Security and Ecological Stability, 147–168.

overshadowed ensuing ecological problems. This was the case when Central Asia was under the auspices of the Soviet Union. When independence was achieved, the situation only became more fragile. While the environmental ramifications of inefficient water use continued, they became even more salient when Kyrgyzstan, Tajikistan, Uzbekistan, Turkmenistan, and Kazakhstan carried on with uncoordinated water use regimes. Thus, while water allocation issues have eclipsed the environmental concerns in the region, solving the water allocation dispute could provide a gateway to solving the broader environmental problems.

After the collapse of the Soviet Union, decisions for managing the two rivers were no longer handed down from Moscow. Each individual republic began either asserting its right to utilize the same amount of water it used during Soviet times, as in the case of Uzbekistan, Kazakhstan, and Turkmenistan (the downstream countries)—or more water, given the new geopolitical realities and the need to develop the national economy, as in the case of Kyrgyzstan and Tajikistan (the upstream countries). In principle, the former position has prevailed. In 1992, the five republics agreed that the water divisions established during Soviet times would continue to be honoured in the future. Acknowledging, however, the uneasiness Kyrgyzstan and Tajikistan demonstrated towards the unsatisfactory status quo, additional agreements were negotiated with Uzbekistan, Kazakhstan, and Turkmenistan. These were barter agreements where fuel, coal, and natural gas were provided by downstream states to their upstream neighbours for reservoir upkeep and an uninterrupted release of water allocations during the cotton-growing season. Despite these arrangements, each of the five republics has expressed dissatisfaction.

Perhaps the most comprehensive regional agreement that recognizes this dissatisfaction is the 1998 Framework Agreement. Given the principles, albeit very general, that are mentioned in the agreement it is apparent that the five republics wish to solve this dispute on terms acceptable to all the states. So, while the core positions of the downstream states seem to be conflicting with those of the upstream states, there is a general agreement that the continuation of the status quo is unacceptable. This general consensus provides an important incentive to forge a new agreement.

This chapter will consider the history of the water dispute and the failures of past barter agreements. In addition, the chapter will provide incite as to how the five republics may be able to achieve a successful formula for resolving their dispute by applying lessons from negotiations among other countries in similar situations. The chapter will review a number of international water agreements that reveal patterns of compromise for conflicting uses between riparian states, which provide a good model for a future Central Asian water agreement. While the respective issues and historical contexts are not exactly the same as in Central Asia, the principles negotiated and the solutions devised in these international water agreements can be useful. Interestingly, one of these agreements is between two of the Central Asian republics, on two smaller rivers.

2. CHALLENGING POSITION AND STRATEGIC LOCATION

A few years after gaining independence, Kyrgyzstan made a number of unilateral decisions that threatened the status quo that had been maintained in the region. In October 1997, Kyrgyz President Askar Akaev signed an edict codifying Kyrgyzstan's right to profit from water resources originating within its territories (Hogan, 2000). In June 2001, Kyrgyzstan adopted a law that classified water as a commodity. In August of the same year, the Kyrgyz government announced that it was preparing regulations to charge neighbouring states, including Kazakhstan and Uzbekistan, for using water (Khamidov, 2001). Clearly intending to follow its plans through, Kyrgyzstan threatened to sell water to China if Uzbekistan refused to pay. In addition, Kyrgyzstan demanded compensation for revenues that were lost because of the release of water downstream to Uzbek farms that could also have been used to generate hydroelectric power (Hogan, 2000). Further destabilizing the situation, Tajikistan has begun to follow the lead of its relatively small and weak, yet water rich neighbour.

The ability of such small and relatively weak nations to contemplate these measures, stem from one advantage: their strategic location in the river basin. Kyrgyzstan and Tajikistan control the headwaters of the Syr Darya and Amu Darya—they are upstream. Furthermore, the two nations also control the main reservoirs and facilities, which release water to their downstream neighbours: Uzbekistan, Kazakhstan, and Turkmenistan. In short, mountainous Kyrgyzstan, along with Tajikistan, are the main "suppliers" of water in Central Asia, while the low-lying states of Kazakhstan, Turkmenistan, and Uzbekistan are the major "consumers" of water—relying on about 90 percent of water coming from outside their borders. According to some estimates, Uzbekistan uses about 3/5 of the regional water supply (Elhance, 1996:212). In contrast, upstream Kyrgyzstan and Tajikistan consume much less water. Since independence from the Soviet Union, both countries have demanded a fairer share of the two rivers.

How does discussion of strategic location, upstream and downstream considerations, inequities in water use, and the issue of Kyrgyzstan's new law of cash for water tie together with lessons and applications for conflict and cooperation in the Aral Sea Basin? The answer may rest in the historical intricacies of the conflict and international precedent.

3. THE ROOTS OF THE DISPUTE

When the Central Asian republics were part of the Soviet bloc, strategic location and control of water mattered little. Decisions regarding when to release water from upstream reservoirs came not from Bishkek and Dushanbe but from Moscow. The Soviet system dictated that Uzbek SSR, Kazak SSR, and Turkmen SSR would grow cotton, while Kyrgyz SSR and Tajik SSR would supply the needed water.[i] Water was regularly released from the Soviet -built reservoirs in spring and summer for agricultural production in the fall. Since Kyrgyz SSR and Tajik SSR could not release this water in the winter to produce hydroelectricity for heating their cities, Moscow directed Uzbek SSR, Kazak SSR, and Turkmen SSR to supply their neighbours with coal and natural gas—basically free of charge. Water in the summer for energy in the winter—the barter system was both dictated and guaranteed from above. Furthermore, Moscow, with its national coffers replete with cotton revenues, contributed capital to the maintenance and upkeep of the reservoirs operated by Kyrgyz SSR and Tajik SSR.

The fall of the Soviet Union changed the geopolitical landscape. The once Soviet - ruled Central Asian republics became independent states with physical borders and dividing lines that transcended more than just their territory. Decisions were no longer made in Moscow but rather in the capital of each individual republic. Of course, cotton continued to be the main "cash crop" of Uzbekistan, Kazakhstan, and Turkmenistan and from their perspective, the life-line for growing it, water, could not come from anywhere else but the reservoirs in Kyrgyzstan and Tajikistan. The same seasonal irrigation scheme had to be employed—water had to be released in the spring and summer and, as argued by the downstream states, in the same quantity. Kyrgyzstan and Tajikistan, however, no longer under the yoke of Soviet orders, did not see the benefit of continuing the old sharing regime. They wished to develop and use their hydropower potential for their own energy needs.

Despite upstream desires, the weight of powerful downstream states triumphed. The five republics met in 1992 and concluded that the old water-sharing regime would stand and the cotton-planting season would not be interrupted (Weinthal, 2002:125).[ii] With no Soviet Union to collect the profits, the cotton revenues would go solely to downstream states, which would continue to consume most of the water. Additional bilateral and some multilateral agreements have governed the water relations among these states since the 1992 agreement. To ameliorate Kyrgyzstan's annoyance with its inability to produce hydropower in the winter, downstream states have agreed to buy Kyrgyz hydroelectricity in the summer when water is released.[iii] In the winter, when Kyrgyzstan refrains from releasing water downstream in the greater interest of cotton growing, Uzbekistan, Kazakhstan, and Turkmenistan provide its upstream neighbour with coal and natural gas for heating and lighting its cities. For its part, Tajikistan has also entered into barter agreements with its downstream neighbours.[iv] Unlike the system of water sharing under the Soviets, when the barter system was that of good for good, without payment, fuel is now sold at market prices or traded for a summer allotment of water. Today, downstream states are able to extract and produce these goods at a relatively low cost. Similarly, the cash once provided by the Soviets to maintain the reservoirs upstream is no longer forthcoming.

3.1. A General Assessment of Central Asian Barter Treaties

If the barter agreements operated efficiently, they would provide a way out of the impasse between the conflicting uses of upstream and downstream states. By engaging in barter arrangements, the five republics have, in essence, linked two major issues—water and energy—that have been a major part of their dispute. Such a 'linkage' strategy expands the range and benefits provided by negotiations across multiple issues (Bennett, Ragland, and Yolles, 1998:66). However, as indicated, these agreements have been far from perfect.

In general, Kyrgyzstan and Tajikistan argue that the water allocation regime modelled on the Soviet system must be revised. Both states use a small portion of the water—they are allowed to withdraw less than 15% of the water—and yet wish to use more than their current allotted quantity for agricultural, industrial, and domestic purposes. Second, the trade-off of free summer water for free winter fuel has been replaced by market prices and complex formulas (Kemelova and Zhalkubaev, 2003:481; Wines, 2002:A14; Weithal, 2002:187-188). The barter system developed has, therefore, been more a source of contention than a source of compromise and cooperation. Take the case of Uzbekistan and Kyrgyzstan, for example. Both countries have cut off their energy deliveries (whether it is coal and natural gas or hydroelectricity) to one another more than once because of outstanding debts (Khamidov, 2001). The republics have also complained that they are either purchasing hydroelectricity they do not need, as in the case of Uzbekistan and Kazakhstan, or buying unnecessary coal and natural gas rather than developing their hydroelectric potential, as in the case of Kyrgyzstan and Tajikistan (Economist, 2003:10; ICG, 2003:15; Klotzli, 1997:422-423). Since most of the water is delivered to Uzbekistan in the summer and stored in the winter, Kyrgyzstan must refrain from producing hydroelectric energy when it really needs it. This is both costly and inefficient, forcing Kyrgyzstan to rely on imported electricity from Uzbekistan to make up for the shortfall (Bransten, 1997). At the same time, the barter agreements are usually delayed till late spring or even early summer—the very time when downstream countries need the water for irrigation. This means that in some years less water than anticipated is delivered in the summer and spring because it is released in the winter for the production of hydroelectricity (Horsman, 2001:75). Had the energy supplies been delivered before the arrival of the warm months, Kyrgyzstan would have had less incentive to produce energy for heat and store more water for the summer (ICG, 2003:14).

The barter system has been subject to other complications. In rainy years, when downstream states irrigate less, they often return less fuel in the winter. In dry years when upstream states release less water they also receive less fuel in return (Wines, 2002:A14). Perhaps, the main source of contention is the fact that Kyrgyzstan and Tajikistan are operating and maintaining the reservoirs on the upper reaches of the Syr Darya and Amu Darya without any financial assistance from Uzbekistan, Kazakhstan, and Turkmenistan. Tajikistan and Kyrgyzstan are, therefore, maintaining a system that benefits others more than themselves (Maynes, 2003:126). Uzbekistan and Kazakhstan have argued that the negotiated barter agreements are a means to supply a tangible good for maintaining the water reservoirs and facilities. According to Kyrgyzstan, however, coal and natural gas can't pay for the costs of facility operation and maintenance. In addition, the value of the bartered goods is often less than the price of facility upkeep (Heltzer, 2003:13). Kyrgyzstan has, therefore, vociferously challenged the barter agreements. It demands some form of cash payment, particularly if the status quo of water deliveries continues to overwhelmingly benefit the lucrative downstream agriculture.

4. INTERNATIONAL WATER LAW, INTERNATIONAL AGREEMENTS AND THE CENTRAL ASIAN WATER DISPUTE

Downstream states have condemned Kyrgyzstan's decision to charge for water. International water law, they argue, does not condone profiting from water resources or charging for shared water.[v] While President Akaev's 1997 edict and Kyrgyzstan's 2001 water law seems to suggest that Kyrgyzstan wishes to exploit her upstream advantage, the new law does not imply payment for actual water. First, Kyrgyzstan's new law is a means to collect payments for the facilities it is operating, which benefit downstream states. Second, it is a scheme put in place to receive compensation for the benefits created downstream by the seasonal water allocations, which are in-turn relinquished upstream. Kyrgyzstan is, therefore, not charging for water *per se*. It is charging for the benefits derived. Yet even with regards to this policy, downstream states have protested. How does international water law address the parties' concerns? Does it provide any solution to the problem?

International law, in general, and international water law, in particular, make the claim that all individuals have a right to water, especially when taking into account the vital needs of humans. This clause was first codified more than 50 years ago in The Universal Declaration of Human Rights—a right to water as a component of the right to life.[vi] The Declaration, however, does not imply in the term "right to water" an unlimited and free right to water. In fact, the 1992 Dublin Statement on Water and Sustainable Development recognizes, in Principle 4, that it is the "basic right of all human beings to have access to clean water and sanitation at an affordable price" and that "past failure to recognize the economic value of water has led to wasteful and environmentally damaging uses of the resource." To that extent, "managing water as an economic good is an important way of achieving efficient and equitable use, and of encouraging conservation and protection of water resources."[vii] While the principles described here reflect on the economic value of water, they provide very little in terms of helping the Central Asian republics out of their impasse.

In terms of more general applications to international water law, Principle 21 of the 1972 Stockholm Declaration claims that "states have in accordance with the Charter of the United Nations and the principles of environmental law, the sovereign right to exploit their own resources pursuant to their own environmental policies, and the responsibility to ensure that activities within their jurisdiction or control do not cause damage to the environment of other states or of areas beyond the limits of national jurisdiction."[viii] Thus, states have a right to utilize their part of the river as they see fit, yet have the responsibility not to harm other basin states given that use. More recently, the 1997 UN Convention on the Law of the Non-Navigational Uses of International Watercourses[ix] states in Article 5 that "...states shall in their respective territories utilize an international watercourse in an equitable and reasonable manner...taking into account the interests of the watercourse states concerned..." Article 7, on the other hand, marks that "...states shall, in utilizing an international watercourse in their territories, take all appropriate measures to prevent the causing of significant harm to other watercourse states." These two articles embody the legal principles of *equitable and reasonable utilization* and the *obligation not to cause significant harm*, respectively. *Equitable and reasonable utilization* has been said to take priority over the *obligation not cause significant harm* (McCaffrey, 2001:308-310) and has also been interpreted to mean that nations, having had a late start in developing their water resources, have a right to an equitable and reasonable utilization of the river's water. Still, those states have the obligation not to cause significant harm to fellow basin states (McCaffrey, 1993:99). Similarly, in determining what is equitable and reasonable, states are to refer to a non-exhaustive list of factors spelled out in Article 6. These range from factors such as 'effects of the use or uses of the watercourses in one state on other watercourse states' to 'existing and potential uses of the watercourse'. These factors are also equal in weight to one another, whereby one does not take priority over the other.

It is clear, that while international legal principles provide hints and suggestions as to what states need to consider in settling disputes over conflicting uses, these legal clauses are rather general, contradictory and vague. Similarly, these clauses are also deficient in prescribing solutions that touch on the issues most pressing to the Central Asian republics, such as compensation pertaining to conflicting utilization plans and side-payments pertaining to benefits created.

Perhaps the most relevant international legal clause to address the issue of payment is Article 25 of the UN Convention. It indicates that "...watercourse states shall participate on an equitable basis in the construction and maintenance or defrayal of the costs of such regulation works as they may have agreed to undertake." The Convention, therefore, affirms a state's obligation to contribute to the costs of water facilities it benefits from, but it falls short of helping states to resolve conflicting claims on a river. Therefore, while international legal principles concur that water is an economic good and recognize that states need to reconcile conflicting uses, there is no explicit clause that determines how such a scheme could be adopted by disputing states sharing a common water body—as the Central Asian case clearly demands.

4.1. International Water Agreements: A Set of Selected Treaties

The aforementioned review of international water law makes clear, that while general legal principles may be used to support differing positions of states, it is in actual negotiations that claims are settled. In fact, a closer look at international water treaties reveals how conflicting uses of a shared river are negotiated. Agreements also provide examples of how the balance between 'equitable utilization' and 'significant harm' is expressed in practice. Specifically, agreements reveal how concepts such as compensation, the reconciliation of conflicting uses, and the utilization of shared waters are formalized. As demonstrated from the discussion on international water law, these concepts are not specifically mentioned in international declarations or legal conventions. This section will consider a set of international water agreement and demonstrate how these concepts are reconciled and negotiated. A close review of these negotiations reveals patterns that can guide future negotiations, particularly over the Syr Darya and Amu Darya.

What are some examples of international water treaties that pertain to concepts and issues that are currently of great contention to the basin states of the Syr Darya and Amu Darya? Table 1 includes a selection of water agreements that span such specific issues as hydropower, flood control and water allocation. Furthermore, the table sorts the treaties according to two key principles that repeat themselves in agreements that pertain to conflicting uses over a shared watercourse. These treaties either pertain to some form of *compensation for facility use* and/or recognize the notion of *compensation for downstream benefits* created. Table 2 provides a more detailed description of these treaties.

Table 1: Selected agreement pertaining to compensation for facility use and downstream benefits

Countries (Upstream and Downstream)	**Compensation for Facility Use**	**Compensation for Downstream Benefits**
Sudan and Eritrea		1925/1951 Gash River Treaty[x]
Nepal and India		1954/1964 Kosi River Agreement[xi]
Nepal and India		1959 Gandak River Agreement[xii]
Canada and United States	1961/1964 Columbia River Agreement[xiii]	1961/1964 Columbia River Agreement
Bhutan and India	1974 Chukha Hydroelectric Agreement on the Wangchhu River[xiv]	1974 Chukha Hydroelectric Agreement on the Wangchhu River
Afghanistan and Iran		1975 Helmand River Agreement[xv]
United States and Canada	1985 Red River Agreement[xvi]	
Canada and United States	1989 Souris River Agreement[xvii]	
Bhutan and India	1995 Kurichhu Hydroelectric Agreement on the Kurichhu River[xviii]	1995 Kurichhu Hydroelectric Agreement on the Kurichhu River
Bhutan and India	1996 Tala Hydroelectric Agreement on the Wangchhu River[xix]	1996 Tala Hydroelectric Agreement on the Wangchhu River
Kyrgyzstan and Kazakhstan	2000 Chu and Talas Rivers Agreement[xx]	

Table 2: A detailed account of a selected number of treaties

Compensation for Facility Use	Compensation for Downstream Benefits
1961/1964 Columbia River Agreement Canada is to build 15,500,000 acre-feet of usable storage in Canadian territory for flood control, and improved water flow and hydropower production. USA is to operate, maintain, and construct hydroelectric facilities in its territory. USA is to pay Canada $64,000,000 for the construction of flood control storage facilities equivalent to 8,450,000 acre-feet of the total usable storage. Compensation is also given for each of the first four flood periods and for operating facilities during those flood periods—$1,875,000. USA will also provide electric power equal to the hydroelectric power lost by Canada as a result of operating the storage to meet flood control functions. Additional, compensation is given to Canada for operating costs and opportunity cost of foregoing alternative uses during flood periods. 1974 Chukha Hydroelectric Agreement on the Wangchhu River India and Bhutan will undertake the Chukha Hydroelectric project to be built in Bhutan for the benefit of both countries. The project shall include a diversion dam, tunnel, a powerhouse with four installation units, and a transmission system to deliver part of the power within Bhutan and to the India-Bhutan border for use by India. The project will be financed by India. 60% of the financing will be in the form of	1925 and 1951 Gash River Agreements Eritrea pledges to use 65 MCM/y. The amount remaining in the river shall go to Sudan. Sudan pays Eritrea each year a share of the sum, which it receives in respect of cultivation by irrigation of land in the Gash delta amounting to 20% of such sum received by Sudan in excess of 50,000 pounds annually. 1954 and 1961 Kosi River Agreement The Kosi project consists of the construction of a barrage, head-works and other appurtenant works in Nepal. The project will provide hydropower, flood control, and irrigation benefits to both countries.[xxi] All works are to be done at the cost of India. Works to be constructed by India are to allow for the irrigation of 93,000 hectares of land in Nepal. Barrage is to be built in Nepal. Nepal is entitled to 50% of the hydroelectric power generated at any power house in a 10 mile radius of the barrage site on payment of such tariff rates as may be fixed for the sale of power by India in consultation with the Nepal. Nepal will receive royalty in respect of power generated and utilized in India at rates to be settled by agreement—no royalty will be paid on the power sold to Nepal. Payment of royalties shall also be provided to Nepal for materials obtained from Nepal to build the barrage. Compensation shall also be provided for land and property used

a grant while 40% will be in the form of a loan (5% interest). Additional funds will be provided as needed by India under the same terms. 1985 Red River Agreement Canada and the US agree to upgrade levee segments on their side of the border and similarly construct an international levee segment along the border. The projects will protect communities on both sides of the border from flooding. Canada shall acquire the lands and construct the levee on her own side of the border. The US shall do the same on her side of the border. Canada is to pay the United States for the construction of the international levy segment. Similarly, Canada shall pay the United Stated $17,000 for routine maintenance of the international levee segment. 1989 Souris River Agreement Canada is to construct Rafferty and Alameda Dams, providing the US with a minimum of 377,800 acre-feet of flood control storage capacity and providing Canada with water supply benefits. The United States is to pay Canada $41.1 million for the flood control storage provided by the Rafferty and Alameda Dams. Canada will operate and maintain the two dams at no cost to the United States. 1995 Kurichhu Hydroelectric Agreement on the Kurichhu River India and Bhutan will undertake the Tala Hydroelectric project to be built in Bhutan for the benefit of both countries. The project shall	for the works and for the loss of land revenue at time of acquisition. India would then own these lands. Coordination committee is set up. In the early to mid 1960s, Nepal demanded that the 1954 Treaty be renegotiated and the 1966 Kosi River Agreement is concluded. Now India does not own the land but rather leases it from Nepal after a payment of compensation. As for the energy entitled to Nepal that would be taken from any powerhouse in India, the Indian Government would construct the necessary transmission line or lines to a point at the Nepal-Indian border. The tariff rates would also be fixed by mutual agreement. 1959 Gandak River Agreement The Gandak project consists of the construction of a barrage, head-works, and other appurtenant works both in Nepal and on the small part of the river that forms the border between the two countries. The project will provide hydropower, flood control and irrigation benefits to both countries.[xxii] India fully funds the works in Nepal. Compensation given for land acquired for works. Royalty on materials needed is also paid to Nepal. Specific navigation rights are provided. Nepal is entitled to irrigation canals to be built by India, at cost to India. These are to be handed over to Nepal whereby operation and maintenance are the responsibility of Nepal. Other channels to be built are to be at the cost of Nepal with India contributing a reasonable sum of money to meet the cost of construction. Hydropower house owned by India but set amount of

include a diversion dam, tunnel, a powerhouse with four installation units, and a transmission system to deliver part of the power within Bhutan and to the India-Bhutan border for use by India. The project will be financed by India. 60% of the financing will be in the form of a grant while 40% will be in the form of a loan (10.75% interest). Additional funds will be provided as needed by India under the same terms. 1996 Tala Hydroelectric Agreement on the Wangchhu River India and Bhutan will undertake the Tala Hydroelectric project to be built in Bhutan for the benefit of both countries. The project shall include a diversion dam, tunnel, a powerhouse with six installation units, and a transmission system to deliver part of the power within Bhutan and to the India-Bhutan border for use by India. The project will be financed by India. 60% of the financing will be in the form of a grant while 40% will be in the form of a loan (9% interest). Additional funds will be provided as needed by India under the same terms. 2000 Chu and Talas Rivers Agreement Kyrgyzstan and Kazakhstan agree that Kyrgyzstan (the party-owner of interstate facilities) is entitled to receive compensation from Kazakhstan (the party-user of interstate facilities) for the costs needed to provide safe and reliable operation.	power sold to Nepal at cost of production. India will build the transmission line so as to supply the power within Nepal. Ownership and management of the hydropower house shall be transferred to Nepal on one year's notice and after the full load of 10,000 KW (60%) had been developed in Nepal. For 15 years, Nepal would sell the power to India at cost. Nepal could elect to purchase the transmission line from India at cost minus depreciation. 1961/1964 Columbia River Agreement Half the downstream power benefits, created by the improved water flow due to the storage dams to be built in Canada, are to be provided to Canada. In the 1964 exchange of notes between Canada and the USA, Canada sold these downstream power benefits to the USA for $254,000,000. 1974 Chukha Hydroelectric Agreement on the Wangchhu River Bhutan is the owner of the project. Bhutan agrees that the surplus power from the project (all the power over and above that required for use in Bhutan) shall be sold to the Government of India. 1975 Helmand River Agreement Iran and Afghanistan agree on a water-sharing regime—specifically water allocations to Iran. Iran offers financial payments and concessional transit rights for Afghan exports through Bandar Abbas in return for more water.[xxiii] 1995 Kurichhu Hydroelectric Agreement on the Kurichhu River Bhutan is the owner of the project.

	Bhutan agrees that the surplus power from the project (all the power over and above that required for use in Bhutan) shall be sold to the Government of India. 1996 Tala Hydroelectric Agreement on the Wangchhu River Bhutan is the owner of the project. Bhutan agrees that the surplus power from the project (all the power over and above that required for use in Bhutan) shall be sold to the Government of India.

As the tables above demonstrate, *compensation for facility use* and *compensation for downstream benefits* are notions repeated in treaties that span such specific issues as flood control, hydropower, and even water allocation. They specifically form a part of agreements negotiated between upstream and downstream countries. International law and international water law establish only general and vague clauses. For example, states have both a right to use a part of the river in their territory but at the same time have an obligation not to harm the other states. In addition, although Article 5 is considered as prevailing over Article 7 in the 1997 UN Convention, it simply entails that states shall take into consideration the interests of other states, incorporating them into their own development plans. It also suggests increased support for reconciling the various interests of river basin states in the development of their shared waters (Wouters, 1997:xxiv; McCaffrey, 1998:22; Eckstein, 2002:85). Therefore, although the main legal clauses seem to entail a compromise, they are still vague, contradictory, and relatively general.

The treaties above demonstrate that the compromise, which international water law has defined so vaguely, is better ascertained from past precedent of international water agreements. To be fair, one specific legal clause in the 1997 UN Convention pertaining to *compensation for facility use* rests in Article 25 and is of great relevance to Central Asia—and the treaties mentioned above embody this principle. In fact, the cases above reveal that states representing different continents and river basins, and economic and political orders have settled their differences in remarkably similar fashions. These same lessons may be applied to Central Asia.

4.3. Applying international precedent to Central Asia

Assessing the Agreements

The agreements reviewed above depict two main principles that are repeated in international water agreements that deal with conflicting uses and interests among states on a given river. In fact, *compensation for facility use* and *compensation for downstream benefits* are principles that capture the essence of the Central Asian water dispute. Following international precedent, these principles could also guide the five republics to a satisfactory resolution to their water dispute. Although these two principles have been separated in the tables above, it is apparent that the two are not mutually exclusive. For example, a state may provide a form of payments for the facilities it is using, which are located upstream, yet implicitly it is also paying for the benefits it is receiving from these facilities. Similarly, the benefits created downstream are a function of facilities and works built upstream, which the downstream state may also fund. Therefore, while the two principles are separated in the tables above, they are also related.

It is also important to highlight some of the differences between the agreements provided above and the situation in Central Asia. In the agreements pertaining to the principle of *compensation for facility use*, the facilities are to be newly built, and the compensation made by the downstream country is a lump sum provided to cover a portion of the costs. It is noteworthy, that the water reservoirs in Kyrgyzstan and Tajikistan, presently under contention, were already built and fully funded by the Soviets. However, two examples—Red River and Columbia River (during flood periods)—point to the notion that a routine payment for operation and maintenance of facilities that benefit another state is not uncommon. In any case, and despite the different payment regimes expressed in the examples above, compensation makes up an important part of any agreement that calls on one party to build or maintain facilities that also benefit another party. This arrangement is, therefore, applicable to Central Asia.

Remarkably, one of the treaty examples above that can be best applied to the dispute over the Syr Darya and Amu Darya is indigenous to Central Asia. On January 21, 2000, Kazakhstan and Kyrgyzstan signed a treaty whereby Kazakhstan agreed to pay, in cash, a share of the operation and maintenance costs of water facilities owned by Kyrgyzstan.[xxiv] These facilities also benefit downstream Kazakhstan and the two countries came up with a cost sharing agreement in which Kyrgyzstan would be compensated for the upkeep of these reservoirs—putting in practice the principle of *compensation for facility use*. Although the agreement referred to facilities and reservoirs located on the Chu and Talas Rivers, a similar strategy could be worked out for the Syr Darya and Amu Darya Rivers and the respective upstream facilities in Kyrgyzstan and Tajikistan.

The agreements presented above dealing with *compensation for downstream benefits* can also be informative for Central Asia. While none of the agreements deal with the trade-off between water for hydropower and water for agricultural production, as is the case in Central Asia, the agreements do deal with benefits accrued downstream from actions taken (or not taken) upstream. In two cases, the Gash River and Helmand River, compensation is provided to the upstream state from the downstream state solely for providing an uninterrupted water allotment downstream. Payment for water is indeed a sensitive issue and may reflect the notion that one state owns the water source and in turn can sell it. Yet, it may also indicate that an upstream state, which gives up use of the waters in its territory for the benefit of the downstream country, shall be compensated.

The other manner by which one can consider these water-sharing agreements is in the context of the treaties that relate to hydropower and flood control. All together, these agreements reflect a notion of sharing and managing a joint river, which exemplifies that when the upstream country takes actions to use and manage the river for the benefit of a downstream state, or provides its strategic upstream territory for the building of infrastructure, the downstream country is to pay reasonable compensation for the benefits accrued to it. In Central Asia, upstream states are constrained not only by their limited water allocations, but also by having to forego uses upstream for benefits downstream.

Here, another agreement indigenous to Central Asia may be instrumental. Although the 1998 Framework Agreement on the Syr Darya (not mentioned in the tables above) is general, it recognizes, at least in principle, the need to provide monetary compensation given the upstream states' foregone benefits, which in-turn favour downstream uses. Specifically, Article 4 and Article 10 make reference to the option that "compensation shall be made…in monetary terms…for annual and multi-year water irrigation storage in the reservoirs," and the "replacement of barter settlements by financial relations," respectively.[xxv] Of course, these monetary and financial regimes have not yet been implemented.

The next section will elaborate more on the possible applications and ramification of a monetary regime that recognizes compensation for facility use and compensation for downstream benefits in Central Asia.

Additional Applications

The section above demonstrates that monetary regimes have been concluded for disputes, which span facilities and works that benefit both countries. Precedent has shown that such a regime is quite common. Assuming that the five republics apply the same principles underscored in the agreements mentioned above to a future agreement on the Syr Darya and Amu Darya, what else can a deal, trading cash for reservoir and facility upkeep—which undoubtedly alleviates an important part of the dispute—do for the conflict as a whole?

In the case of the Aral Sea Basin, one notable benefit would be efficiency in water use. Just as water pricing, in general, is used to promote efficiency, an established system of trading cash for facility operation and maintenance (according to water used) would undoubtedly force users of water to think twice as to how they will utilize the water they receive. Currently, cotton production in downstream states uses water in an extremely inefficient manner. According to some estimates, irrigation efficiencies are said to be no better than 40% or 50% (UNDP, 1995:5). Undoubtedly, this is related to out-of-date and inefficient irrigation facilities and schemes. However, these inefficiencies can also be attributed to the inequitable distribution of water among the states according to outdated Soviet principles and the lack of an efficient pricing scheme for water. A true valuation of water would give downstream governments an incentive to charge farmers a price more reflective of the value of the good and farmers more incentive to use water more efficiently. A restructuring of the water-sharing regime is also necessary, given that the upstream states have continued to be short-changed since independence from the Soviet Union. A compensation scheme could again prove instrumental. If downstream farmers use less water for agriculture, more water is available for the benefit of upstream states. Water used more efficiently would also have a positive affect on the Aral Sea, with more water flowing downstream rather than being lost to stretches of wastewater or marshlands.

What about the hydropower generation, which upstream states are arguing they are being denied from developing, given the agricultural needs of downstream states?

Since cotton irrigation must take place on a rigorous seasonal schedule, whereby water is to be released from upstream reservoirs in the spring through summer, efforts to develop hydropower in the winter seem to contradict agricultural water use. Here, the concept of *compensation for downstream benefits* can be employed. Downstream states receive great benefits from upstream facilities and reservoirs and from upstream efforts to refrain from releasing water in the winter for hydropower development. According to the International Crisis Group, Kyrgyzstan has already indicated that it will be willing to cease producing electricity at the Toktogul reservoir during the winter, as a long-term solution, if downstream states compensate for the losses (ICG, 2002:16-17). In fact, the downstream states are some of the world's greatest producers and exporters of cotton and this is largely due to the trade-off currently being made upstream. A percentage of the cotton profits paid to upstream states could provide the appropriate compensation for the loss of hydropower generation. Upstream states can, therefore, share in the downstream profits derived from water used for economic development (Glantz, 1999:23). This notion, while not specifically mentioned in the 1998 Framework Agreement, can be attributed to the spirit and substance of both Articles 4 and 10. It also mirrors international precedent, whereby actions taken (or not taken) upstream, which benefit downstream states, are to be recognized through compensation.

5. CONCLUDING REMARKS

The 2000 Chu and Talas Rivers Agreement and the 1998 Framework Agreement have signalled that the Central Asian republics have come closer to reconciling their main differences. Yet the continued conflict between upstream and downstream countries over the Syr Darya and Amu Darya Rivers demonstrates that the notion of compensation for water is still a sensitive one. Nonetheless, if it is recognised that the idea implies compensation for facility use and accrued benefits, negative perceptions regarding this proposal could be reduced. Citing other international examples and agreements that have articulated these principles is even more valuable—demonstrating how international precedent could be applied in Central Asia.

Kyrgyzstan's challenge to its neighbours should, therefore, be considered not as grounds for intensified conflict but rather a key to the solution of the Central Asian water dispute and the Aral Sea environmental tragedy. Similarly Kyrgyzstan's, and subsequently Tajikistan's, claims are based on two notions that have been negotiated in international agreements. It is also important to understand that by reverting to a fair and reasonable compensation regime dominated by monetary and financial dealings rather than the failed linkage regime, downstream states would be able to gain more control over the water supplies flowing downstream. They would be able to demand appropriate services from the Kyrgyz and Tajiks in managing the reservoirs and at the same time continue to benefit from the steady cotton profits. Upstream states will have an incentive and obligation to care for the reservoirs, given the compensation they are receiving. Similarly, they will also commit to releasing the appropriate amounts of water according to the set seasonal schedule, given that they are fairly and appropriately compensated for not producing hydropower in the winter.

The pricing and compensation regime may then produce an "invisible hand effect." With a higher value now attributed to water, downstream governments will have an incentive to promote efficiency among cotton growing farmers, which could in turn lead to a domestic pricing scheme for water. Higher efficiency and less water wasted downstream could also mean additional water flowing into the Aral Sea and perhaps even a different regional water regime, allocating more water upstream.

The agenda discussed here does not pretend to solve the Central Asian water conflict on its own. It will especially not be enough to solve the Aral tragedy, with all its environmental complications, that has grappled the region for decades. However, it does outline and support a strategy linking negotiated principles based on international precedent, which could motivate the parties to a satisfactory resolution to their water dispute, with environmental ramifications for the Aral Sea. Naturally, the cooperation of all five republics in bringing forth responsible and practical proposals to the negotiating table, with the well being of the region as a whole in mind, will be key to the bright future of Central Asia.

REFERENCES

1. Abidi, A. "Irano-Afghan Dispute over the Helmand Water," *International Studies*, V 16, N 3, July-September, 1977.
2. Agreement between the Government of India and the Government of Nepal on the Kosi Project, 25 April 1954; Amended Agreement Between His Majesty's Government of Nepal and the Government of India, 19 December 1966; http://ocid.nacse.org/qml/research/tfdd/toTFDDdocs/85ENG.htm; http://ocid.nacse.org/cgi-bin/qml/tfdd/treaties.qml?qml_screen=full&TN=136
3. Agreement Between His Majesty's Government of Nepal and the Government of India on the Gandak Irrigation and Power Project, 4 December 1959; http://ocid.nacse.org/qml/research/tfdd/toTFDDdocs/111ENG.htm
4. Agreement Between the Government of India and the Royal Government of Bhutan Regarding the Chukha Hydro-Electric Project, 23 March 1974.

5. Agreement Between the Government of India and the Royal Government of Bhutan Regarding the Kurichhu Hydro-Electric Project, 1995. Exact date not known; http://www.bifa.org.bt/Projects.html

6. Agreement Between the Government of India and the Royal Government of Bhutan Regarding the Wangchhu Hydro-Electric Project, 1996. Exact date not known; http://www.bifa.org.bt/Projects.html

7. Agreement between the Government of Canada and the Government of the United States of America for water supply and flood control in the Souris River Basin, 26 October 1989; http://ocid.nacse.org/qml/research/tfdd/toTFDDdocs/181ENG.htm

8. Agreement Between the Republic of Kazakhstan, the Republic of Kyrgyzstan, the Republic of Uzbekistan, the Republic of Tajikistan and Turkmenistan on Cooperation in the Field of Joint Water Resources Management and Conservation of Interstate Sources, 18 February 1992; http://www.ce.utexas.edu/prof/mckinney/papers/aral/agreements/ICWC-Feb18-1992.pdf

9. Agreement Between the Government of the Republic of Tajikistan and the Government of the Republic of Uzbekistan on Cooperation in the Area of Rational Water and Energy Uses, 4 February 1998; http://www.ce.utexas.edu/prof/mckinney/papers/aral/agreements/Kayrakum-98.pdf

10. Agreement between the Governments of the Republic of Kazakhstan, the Kyrgyz Republic, and the Republic of Uzbekistan on the Use of Water and Energy Resources of the Syr Darya Basin, 17 March 1998; http://www.ce.utexas.edu/prof/mckinney/papers/aral/agreements/SyrDaryaAgr-Mar17-98.pdf

11. Agreement Between the Government of the Republic of Kazakhstan, the Government of the Kyrgyz Republic and the Government of the Republic of Uzbekistan on Joint and Complex Use Water and Energy Resources of the Naryn Syr Darya Cascade Reservoirs in 1998, 17 March 1998; http://www.ce.utexas.edu/prof/mckinney/papers/aral/agreements/Annual-Operation-98.pdf

12. Agreement Between the Government of the Republic of Uzbekistan and the Government of the Republic of Tajikistan on Cooperation in the Area of Rational Water and Energy Uses in 1999, 13 April 1999; http://www.ce.utexas.edu/prof/mckinney/papers/aral/agreements/Kayrakum-99.pdf

13. Agreement Between the Government of the Republic of Kazakhstan and the Government of the Kyrgyz Republic on Comprehensive Use of Water and Energy Resources of the Naryn Syr Darya Cascade Reservoirs in 1999, 22 May 1999; http://www.ce.utexas.edu/prof/mckinney/papers/aral/agreements/Annual-KzKg-99.pdf

14. Agreement Between the Government of the Republic of Uzbekistan and the Government of the Republic of Tajikistan on Cooperation in the Area of Rational Water and Energy Uses in 2000, 14 January 2000; http://www.ce.utexas.edu/prof/mckinney/papers/aral/agreements/Kayrakum-00.pdf

15. Agreement between the Government of the Republic of Kazakhstan and the Government of the Kyrgyz Republic on Utilization of the Water Facilities of Interstate Use on the Chu and Talas Rivers, 21 January 2000; http://www.ce.utexas.edu/prof/mckinney/papers/aral/agreements/OandM-Agreement.pdf

16. Agreement Between the Government of the Republic of Kazakhstan and the Government of the Kyrgyz Republic On the Use of Water and Energy Resources of the Naryn Syr Darya Cascade of Reservoirs in 2000, 23 May 2000;

http://www.ce.utexas.edu/prof/mckinney/papers/aral/agreements/Annual-KzKg-00.pdf

17. "At the Crossroads: A Survey of Central Asia," *The Economist*, July 26, 2003, pgs. 10-11.

18. Bennett, L., Ragland, S. and Yolles, P. (1998) "Facilitating International Agreements Through an Interconnected Game Approach: The Case of River Basins," in Just, R. and Netanyahu, S. (Eds.) *Conflict and Cooperation on Transboundary Water Resources*, (Kluwer Academic Publishers, Boston).

19. Bransten, J. "Kyrgyzstan/Uzbekistan: The Politics of Water," *Radio Free Europe/Radio Liberty*, October 14, 1997; http://www.rferl.org/nca/features/1997/10/F.RU.971014134935.html

20. Eckstein, G. (2002) "Development of international water law and the UN Watercourse Convention," in Turton, A. and Henwood, R. (Eds.): *Hydropolitics in the Developing World: A Southern African Perspective*, (AWIRU, Pretoria, South Africa).

21. Elhance, A. "Conflict and Cooperation Over Water in the Aral Sea Basin," *Studies in Conflict and Terrorism*, V 20 (1997).

22. Exchange of Notes between the United Kingdom and Italy respecting the Regulation of the Utilization of the Waters of the River Gash, 12/15 June 1925; Exchange of Letters between the irrigation adviser and director of irrigation, Sudan government, and the controller of agriculture, Eritrea, 18 April 1951; http://ocid.nacse.org/qml/research/tfdd/toTFDDdocs/33ENG.htm

23. Exchange of Notes between the Government of Canada and the Government of the United States of America constituting an Agreement concerning the Construction of a Joint Ring Levee—Red River, 29 and 30 August 1988; http://ocid.nacse.org/qml/research/tfdd/toTFDDdocs/248ENG.htm

24. Glantz, M. (1999) "Sustainable Development and Creeping Environmental problems in the Aral Sea Region," in Glantz, M. (Ed.) *Creeping Environmental Problems and Sustainable Development in the Aral Sea Basin*, (Cambridge University Press, Cambridge, UK).

25. Helmand River Water Treaty, 13 March 1973.

26. Heltzer, G. "Stalemate in the Aral Sea Basin: Will Kyrgyzstan's New Water Law Bring Downstream Nations Back to the Multilateral Table?" *Georgetown International Environmental Law Review*, V 19, N 291, (2003).

27. Hogan, B. "Central Asian States Wrangle Over Water," Eurasinet.org. April 5, 2000; http://www.eurasianet.org/departments/environment/articles/eav040500.shtml

28. Horsman, S. (2001) "Water in Central Asia: Regional Cooperation or Conflict," in Allison, R. and Jonson, L. (Eds.) *Central Asian Security: The New International Context*, (Royal Institute of International Affairs and Brookings Institution Press, London and Washington, D.C.).

29. Intergovernmental Protocol Between the Government of the Kyrgyz Republic and the Government of the Republic of Uzbekistan on Use of the Naryn Syr Darya Water and Energy Resources in 2000, 16 March 2000; http://www.ce.utexas.edu/prof/mckinney/papers/aral/agreements/Annual-KzKg-00.pdf

30. International Crisis Group, "Central Asia: Water and Conflict," *Asia Report*, N 34, May 30, 2002.

31. Kemelova, D. and Zhalkubaev, G. "Water, Conflict, and Regional Security in Central Asia Revisited," *New York University School of Law Environmental Law Journal*, V 11, N 479, (2003).

32. Khamidov, A. "Water Continues to be Source of Tension in Central Asia," *Eurasianet.org*, October 23, 2001; http://www.eurassianet.org/departments/environment/articles/eav102301.html

33. Klotzli, S. (1997) "The Aral Sea Syndrome and Regional Cooperation in Central Asia: Opportunity or Obstacle," in Gleditsch, N.P. et al. (Eds.) *Conflict and the Environment*, (Kluwer Academic Publishers, Boston).

34. "Kyrgyzstan to Press Ahead With Plans to Sell Water," Eurasianet.org; http://eurasianet.org/resource/kyrgyzstan/200108/0024.html

35. Maynes, C.W. "America Discovers Central Asia," *Foreign Affairs*, V 82, N 2, (March/April 2002).

36. McCaffrey, S. (1993) "Water, politics, and international law," in Gleick, P. (Ed.) *Water in Crisis: A Guide to the World's Fresh Water Resources*, (Oxford University Press, New York).

37. McCaffrey, S. (1998) 'The UN convention on the law of non-navigational uses of international watercourses: prospects and pitfalls', in Salman. S and Boisson de Chazournes, L. (Eds.): *International Watercourses: Enhancing Cooperation and Managing Conflict*, Proceedings of World Bank Seminar, World Bank Technical Paper, No. 414, July.

38. McCaffrey, S. *The Law of International Watercourses: Non-Navigational Uses*, (Oxford University Press, New York: 2001).

39. Salman, S. and Uprety, K. *Conflict and Cooperation on South Asia's International Rivers: A Legal Perspective*, (The World Bank, Washington, D.C.: 2002).

40. The Dublin Statement on Water and Sustainable Development, 31 January 1992; http://www.wmo.ch/web/homs/documents/english/icwedece.html

41. The Stockholm Declaration on the Human Environment, 16 June 1972; http://www.unep.org/Documents/?DocumentID=97&ArticleID=1503

42. Treaty relating to cooperative development of the water resources of the Columbia River Basin (with Annexes) Done 17 January 1961; Exchange of Notes Constituting an Agreement between Canada and the United States Authorizing the Canadian Entitlement Purchase Agreement Provided for Under the Treaty Relating to Co-Operative Development of the Water Resources of the Columbia River Basin, Done in Washington (17 January 1961), 16 September 1964; http://ocid.nacse.org/qml/research/tfdd/toTFDDdocs/116ENG.htm; http://ocid.nacse.org/qml/research/tfdd/toTFDDdocs/125ENG.htm

43. UNDP, "The Aral Crisis," Tashkent 1995.

44. United Nations Convention on the Law of the Non-Navigational Uses of International Watercourses, 21 May 1997; http://www.un.org/law/ilc/texts/nnavfra.htm

45. Universal Declaration on Human Rights, 10 December 1948; http://www.un.org/Overview/rights.html

46. Weinthal, E. (2002) *State Making and Environmental Cooperation: Linking Domestic and International Politics in Central Asia*, (The MIT Press, Cambridge, Massachusetts).

47. Wines, M. "Grand Soviet Scheme for Sharing Water in Central Asia is Foundering," *New York Times*, December 9, 2002.

48. Wouters, P. (1997) 'Editor's forward: Current status of international water law', in Wouters, P. (Ed.): *International Water Law: Selected Writings of Professor Charles B. Bourne*, (Kluwer Law International, Boston).

49. Zonn, I. (1999) "The impact of political ideology on creeping environmental changes in the Aral Sea Basin," in Glantz, M. (Ed.) *Creeping Environmental Problems and Sustainable Development in the Aral Sea Basin*, (Cambridge University Press, Cambridge, UK).

---

[i] For a short and concise account of the cotton and irrigation regimes dictated by the USSR see Zonn (1999:157-181).

[ii] Agreement Between the Republic of Kazakhstan, the Republic of Kyrgyzstan, the Republic of Uzbekistan, the Republic of Tajikistan and Turkmenistan on Cooperation in the Field of Joint Water Resources Management and Conservation of Interstate Source, 18 February 1992, Article 3.

[iii] Agreement Between the Government of the Republic of Kazakhstan, the Government of the Kyrgyz Republic and the Government of the Republic of Uzbekistan on Joint and Complex Use Water and Energy Resources of the Naryn Syr Darya Cascade Reservoirs, 17 March 1998; Agreement Between the Government of the Republic of Kazakhstan and the Government of the Kyrgyz Republic on Comprehensive Use of Water and Energy Resources of the Naryn Syr Darya Cascade Reservoirs in 1999, 22 May 1999; Agreement Between the Government of the Republic of Kazakhstan and the Government of the Kyrgyz Republic On the Use of Water and Energy Resources of the Naryn Syr Darya Cascade of Reservoirs in 2000, 23 May 2000; Intergovernmental Protocol Between the Government of the Kyrgyz Republic and the Government of the Republic of Uzbekistan on Use of the Naryn Syr Darya Water and Energy Resources in 2000, 16 March 2000.

[iv] Agreement Between the Government of the Republic of Tajikistan and the Government of the Republic of Uzbekistan on Cooperation in the Area of Rational Water and Energy Uses, 4 February 1998; Agreement Between the Government of the Republic of Uzbekistan and the Government of the Republic of Tajikistan on Cooperation in the Area of Rational Water and Energy Uses in 1999, 13 April 1999; Agreement Between the Government of the Republic of Uzbekistan and the Government of the Republic of Tajikistan on Cooperation in the Area of Rational Water and Energy Uses in 2000, 14 January 2000.

[v] Comments made by Nursultan Nazarbaev, July 4, 2001 in "Kyrgyzstan to Press Ahead With Plans to Sell Water," (2001); and comments by Uzbek Prime Minister Utkir Sultanov, October 16, 2001 in Khadimov (2001).

[vi] Universal Declaration of Human Rights, 10 December 1948, Article 3.

[vii] The Dublin Statement on Water and Sustainable Development, 31 January 1992, Principle 4.

[viii] The Stockholm Declaration on the Human Environment, 16 June 1972, Principle 21.

[ix] United Nations Convention on the Law of the Non-Navigational Uses of International Watercourses, 21 May 1997.

[x] Exchange of Notes between the United Kingdom and Italy respecting the Regulation of the Utilization of the Waters of the River Gash, 12/15 June 1925; Exchange of Letters between the irrigation adviser and director of irrigation, Sudan government, and the controller of agriculture, Eritrea, 18 April 1951.

[xi]Agreement between the Government of India and the Government of Nepal on the Kosi Project, 25 April 1954; Amended Agreement Between His Majesty's Government of Nepal and the Government of India, 19 December 1966.
[xii] Agreement Between His Majesty's Government of Nepal and the Government of India on the Gandak Irrigation and Power Project, 4 December 1959.
[xiii]Treaty relating to cooperative development of the water resources of the Columbia River Basin (with Annexes) Done 17 January 1961; Exchange of Notes Constituting an Agreement between Canada and the United States Authorizing the Canadian Entitlement Purchase Agreement Provided for Under the Treaty Relating to Co-Operative Development of the Water Resources of the Columbia River Basin, Done in Washington (17 January 1961), 16 September 1964.
[xiv] Agreement Between the Government of India and the Royal Government of Bhutan Regarding the Chukha Hydro-Electric Project, 23 March 1974.
[xv] Helmand River Water Treaty, 13 March 1973.
[xvi] Agreement concerning cooperation in flood control, with construction plan—Red River, 29 and 30 August 1988.
[xvii] Agreement between the Government of Canada and the Government of the United States of America for water supply and flood control in the Souris River Basin, 26 October 1989.
[xviii] Agreement Between the Government of India and the Royal Government of Bhutan Regarding the Kurichhu Hydro-Electric Project, 1995. Exact date not known.
[xix] Agreement Between the Government of India and the Royal Government of Bhutan Regarding the Wangchhu Hydro-Electric Project, 1996. Exact date not known.
[xx] Agreement between the Government of the Republic of Kazakhstan and the Government of the Kyrgyz Republic on Utilization of the Water Facilities of Interstate Use on the Chu and Talas Rivers, 21 January 2000.
[xxi] For an analysis of the Kosi Agreement see Salman and Uprety (2002:65-82).
[xxii] For an analysis of the Gandak Agreement see Salman and Uprety (2002:83-95).
[xxiii] A description of the treaty is available in Abidi (1977:371). Abidi discusses the payment incentives that were discussed.
[xxiv]Agreement between the Government of the Republic of Kazakhstan and the Government of the Kyrgyz Republic on Utilization of the Water Facilities of Interstate Use on the Chu and Talas Rivers, 21 January 2000, Article 3 and 4; According to the International Crisis Group (2002:16) the Kazakhs have agreed to pay Kyrgyzstan about $100,000 a year for the maintenance of the facilities.
[xxv] Agreement between the Governments of the Republic of Kazakhstan, the Kyrgyz Republic, and the Republic of Uzbekistan on the Use of Water and Energy Resources of the Syr Darya Basin, 17 March 1998, Articles 4 and 10.

CRITICAL GEOGRAPHY – THE STRATEGIC INFLUENCE OF WATER IN CENTRAL ASIA

LIOTTA, P.H.
Jerome E. Levy Chair of Economic Geography and National Security, United States Naval War College

ABSTRACT

This chapter assesses the geostrategic and geopolitical implications of water shortages in Central Asia and argues that resource scarcity or competition cannot be separated from other regional realities. On-again, off-again relations between Central Asian states, as well as the so-called war on terrorism, bear significant implications for the region. Thus, the convention of defining Central Asia as a grouping of five states is increasingly less relevant for policy making and sound strategic analysis. Central Asia is linked to the entire Caspian basin, the South Caucasus, Afghanistan, Pakistan, Iran, Turkey, and China's Xinjiang province. Equally, the long-term interests of Russia and the United States now play into the complex realities of Central Asia. Specifically regarding water shortages, a number of vulnerability issues involving so-called "non-traditional" security present serious long-term challenges to the stability of the region. The chapter argues that there are crucial differences between threats and vulnerabilities, distinguishes between the two, and suggests relevant policy applications for the Central Asian states. The analysis includes a review of theoretical models that have been proposed in research. Specifically, this review addresses what have been argued as "trigger mechanisms" that can unleash violent conflict, create socioeconomic disparity, and induce long-term insecurity, and provides possible pathways for geostrategic solutions and means to reduce water resource tensions.

"Water is a trigger for conflict but a reason to make peace."

Leif Ohlsson[1]

In late August 2003, while a NATO advanced research workshop focusing on trans boundary water resources in Central Asia was underway in Novosibirsk-Akademgorodok, Siberia, a similar United Nations session was taking place in Dushanbe, Tajikistan. Tajik president Imomali Rakhmonov, who had earlier persuaded the United Nations to declare 2003 the "year of fresh water," hosted the conference itself. Rakhmonov, declaring alarm over the ecological destruction of the Aral Sea, which has caused massive internal displacement and today threatens three million inhabitants, also noted that one out of six on the planet today already lacks access to safe drinking water.[2] If true, this outcome would seem to confirm early projections made in the National Intelligence Council's *Global*

H. Vogtmann and N. Dobretsov (eds.), Transboundary Water Resources: Strategies for Regional Security and Ecological Stability, 169–193.

Trends 2015: A Dialogue about the Future with Nongovernmental Experts that by 2015 one out of two on the planet would live in water stressed areas, and by 2023 it would be possible that available water would hit "carrying capacity" and be unable to provide available water to all populations.[3]

What remains most significant about the Dushanbe and Siberia sessions focusing on strategies for regional security and ecological stability, however, remains a widely unacknowledged but largely unspoken truth: *any strategic implementation plan that attempts to solve water issues in Central Asia as a technical, or engineering, problem without recognizing social, political, and economic aspects is bound to fail.* This became evident, in particular, when the NATO workshop failed to incorporate some of these aspects in its final series of recommendations; worse, the UN session in Dushanbe erupted in bouts of mutual antagonism and thinly veiled criticisms of delegates blaming one another for unequal water resource management and continued dependence on Soviet-era plans.[4]

Doubtless, environmental and human security remain both evolving and contested concepts. Yet the *vulnerability* aspects that these security issues involve present serious long-term challenges to the success and stability of Central Asia. This chapter, aside from offering a general approach to the meaning of environmental security, argues that there are crucial differences between *threats* and *vulnerabilities*, distinguishes between the two, and suggests relevant policy applications. The analysis includes a review of theoretical models that have been proposed in recent research and considers their relevance to the region. Specifically, this review addresses what have been argued as "trigger mechanisms" that can unleash violent conflict, create socio-economic disparity, and induce long-term insecurity. Finally, a number of policy issues relevant to future research are considered as bases for stabilizing and supporting foundations during periods of future change and potential crisis.

To be sure, such Soviet-era plans, both past and potentially future, produced and might still produce disastrous consequences for the region. While Communist ideology might reasonably have made each Central Asian state mutually dependent on a relatively equal basis during the time of the USSR (to include barter and water management schemes between republics), postindependence has created a radical dichotomy. More powerful and economically viable "downstream" states—Kazakhstan, Uzbekistan, and Turkmenistan—are all heavy consumers of water, particularly for cotton growing (often in arid, semi-desert environments).[5] "Upstream" states—Kyrgyzstan and Tajikistan— have little natural resources *other* than water. The two main river systems of Central Asia—the Syr Darya from Kyrgyzstan, flowing through Uzbekistan and Kazakhstan; and the Amu Darya, flowing through Turkmenistan and Uzbekistan—now constitute a critical geography, and their specific flows and usage determine economic potential or disaster, social integration or fragmentation, and geopolitical order or chaos.[6]

In the past, Soviet plans intended to solve water issues through sometimes astounding technological undertakings. Yet these attempts, if implemented, would have done little to encourage, or force, proper water management in the Aral Sea basin. One of the more ambitious plans, first tabled in the 1980s, for example, called for the detonation of nuclear weapons in mountainous Central Asia (in Kyrgyzstan and Tajikistan), in theory melting glaciers and allowing water to flow downstream potentially to refill the Aral Sea. (Little, of course, was said at the time about the inevitable human toll in flooding and ecological instability, to include the major loss of a "natural" resource for two Central Asian states.) Another idea that resurfaced in 2001 and continues in some circles as a viable proposition is to divert water from the Ob (which flows from the Arctic) and Irtysh Rivers through massive pipelines to Kazakhstan and the Aral Sea basin. These ideas reflect in precise terms the Marxist view of nature as something to be dominated, not accommodated. Thus, phenomenal expenditures made during Soviet times, while building dams and canals across Central Asia, were bent always on the increasing irrigation potential and did not consider the consequence of environmental damage, let alone the critical concept of ecological balance.

In some Central Asian locations, therefore, only one crop grew before farmers were forced to abandon the land because of salination.[7] Today, the Aral Sea is a permanent "Dead Sea": once the fourth largest lake in the world, yet today a toxic wasteland where salt and dust storms regularly whip up, already forcing the removal of two million inhabitants of autonomous Karalkapakstan within Uzbekistan. To make matters worse, the centralized Soviet management of water led to little discipline in its use, and was fixed by water quotas from Moscow, which largely favored downstream states. (The growing of cotton, or "white gold" as it is referred to in Uzbekistan, exploded during the U.S. Civil War, when the North's tight trade blockade on the South forced Russian textile manufacturers to look to Central Asia for fertile growing areas—relying on the seemingly unlimited supplies of irrigation water from the Aral Sea—and the Czar's empire inevitably expanded to assist the industrialization of modern Russia.[8]) According to Duishen Mamatkanov, Director of the Institute of Water Problems and Hydropower of the Kyrgyz Academy of Sciences, the Soviet Union restricted agriculture in upstream states Kyrgyzstan and Tajikistan to maximize cotton output in downstream states.[9]

Kyrgyzstan and Tajikistan, having only limited natural resources other than water, thus sought to develop hydroelectric potential, which proved incompatible with ensuring sufficient reservoirs for downstream irrigation in spring and summer; to satisfy their energy needs the upstream states received vast quantities of Kazakh coal, gas, and *mazut* (a heavy oil left from refinery residues). In return, Kyrgyzstan and Tajikistan also received electricity from the downstream states during the winter and Moscow covered the immense costs of operating and maintaining dams, reservoirs, canals, and irrigation pump systems.[10]

Notwithstanding the complexities of the Soviet water management and barter system, none of the five independent Central Asian states respects the "dependency" rule set by which they lived under the Soviet system. Upstream states cannot today afford energy resources set by downstream states at realistic world prices, and therefore feel compelled to increasingly generate their hydroelectric energy. Yet Uzbek citizens in the Ferghana Valley (home to radical extremists such as the Islamic Movement of Uzbekistan, which trained with Al Qaeda and fought alongside the Taliban, and remains a forceful presence even today in the region) have endured winter floods when Kyrgyzstan releases dam water to generate electricity and also experienced (increasingly frequent) summer droughts partially due to Kyrgyzstan's increased electricity production on the Syr Darya. In retaliation, Uzbekistan has failed to deliver gas due (under the previous Soviet barter system) in exchange for water. Thus, we see scenarios in which Kyrgyz citizens endure winter periods without electricity and Uzbeks endure both winter floods and summer droughts. Clearly, the bases for mutual cooperation seem to be deteriorating, and mistrust growing.

The ultimate, and unnecessary, irony of all this is that Central Asia does *not* suffer from water shortages. Rather, decaying infrastructure and woefully inadequate irrigation systems, where up to 70 percent of water is lost through evaporation and filtration, as well as failure to reach mutually beneficial agreements for upstream and downstream states, is leading Central Asia in the wrong direction. Further, water usage rates, according to some authorities, are now150 percent beyond recommended levels.[11] The ultimate conclusion from this emerging complexity, therefore, suggests that the geopolitical state of affairs in Central Asia is not good.

Consequently, the post-independence dilemma of both upstream and downstream Central Asian states points to a peculiar paradox: according to Leif Ohlsson, who has studied the Malthusian potentialities of resource scarcity and environmental conflict, disputes over water become more intense the *smaller* the scale of the dispute. Indeed, this hypothesis has largely been strongly supported in the region, where aggression has been localized rather than cast on a wide strategic landscape of all out war. Schematically, Ohlsson has usefully cast this dynamic as a tension between interstate and intrastate competition and conflict:[12]

Types of Water Conflicts	**Supply Increase Attempts**	**Demand Management Attempts**
Interstate Conflicts	1) Attempts to increase supply	2) Attempts to manage demand
Intrastate Conflicts	3) Internal competition between population sectors	4) Water scarcity necessitates demand management

Table 1. Causes of Water Conflicts

Although Ohlsson's argument that we will increasingly see focus shifts (from supply increase attempts between states to water scarcity management attempts states) may be debatable, his observations that the risk of violence becomes greater the smaller the scale of the dispute does have some merit in the Central Asia discussion.[13]

What is happening in Central Asia, whether one considers it the inevitable disorder that rises out of the collapse of an empire or the "morbid interregnum" before the reestablishment of a new and sustaining order,[14] has implications far beyond the region itself. In essence, a grand experiment is taking place in Central Asia. Though it is not yet clear that this experiment is doomed to failure, the morbid symptoms have appeared—and water is a critical uncertainty for the future.

Thus, policy makers should recognize that there is a fundamental linkage between resource issues and security, and that these issues and the long-term future of the region will negatively degrade unless some effort is made to reverse the tide.[15] Identifying and acting on problems of environmental degradation and resource scarcity, in the best possible world, ought to be a common feature of future security policy.

Above all, it remains crucial to recognize that such issues cannot be separated from larger regional cooperative and competitive interests, which also cannot be considered in isolation from each other. Before considering aspects of these security features, nevertheless, it would be worth addressing specific truths about the region itself.

1. THE FIVE REALITIES OF CENTRAL ASIA

The overarching reality of Central Asia today is perhaps so obvious that it is frequently overlooked. No one has phrased the emerging recognition of this reality, however, better than Rajan Menon in observing that:

> The convention of defining Central Asia as a grouping of five states is of diminishing value for effective policy making and sound strategic analysis. A seamless web connects Central Asia proper, the South Caucasus, Afghanistan, Pakistan, Iran, Turkey, and China's

> Xinjiang province. Thinking in terms of a "greater Central Asia" captures the bigger picture and reflects how forces in one part of greater Central Asia will affect its other parts . . . [and] area specialists and theorists may have to co-operate if they are to be of use to practitioners.[16]

If one were to broadly categorize the emerging truths about Central Asia, they might be represented as:

1) The "Stans" are not the "Stans."

Among policy makers in the United States, in particular, it has become quite popular to identify Central Asia in discursive shorthand as the "Stans." Nothing, of course, could be farther from the truth. As the graphic below illustrates, the peoples and the states of Central Asia are anything *but* commonly linked in one overarching, though comfortably identifiable, reality.[17] "Stan," of course, is taken from the Persian, meaning "the land of . . ."

What is evident, nonetheless, is that Central Asian states are not, as the graphic illustrates, the designated lands of specific ethnicities. Although the situation is reversed today, Kazakhstan actually had more ethnic Russians than Kazakhs within its borders at the time of its independence in 1991. Equally, one is likely to find numerous Tajiks in the ancient cities of Samarkand and Bokhara inside Uzbekistan. Part of this mix across borders was due to Stalin's forced movements during his reign (such as the forced removal of all Chechens from Chechnya).

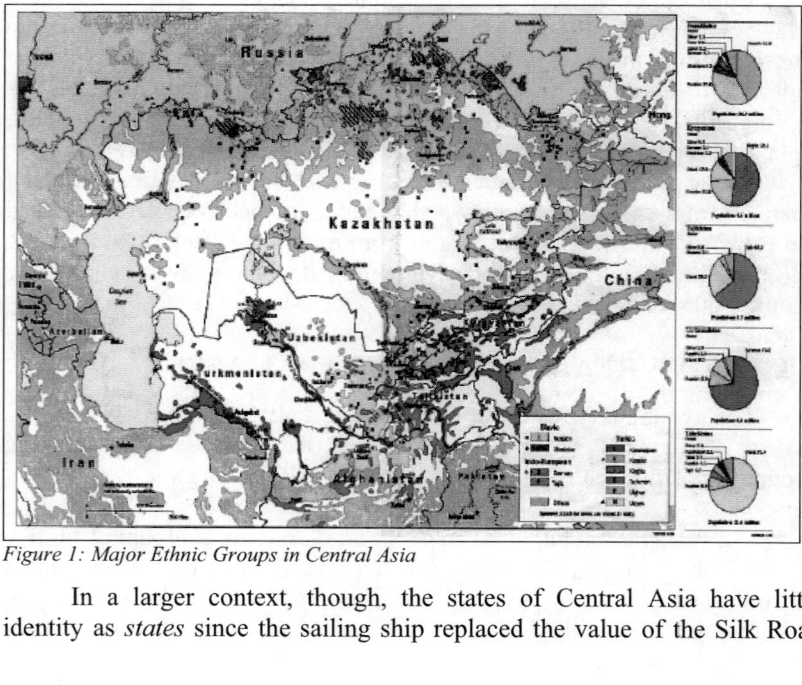

Figure 1: Major Ethnic Groups in Central Asia

In a larger context, though, the states of Central Asia have little identity as *states* since the sailing ship replaced the value of the Silk Road as a trading route half a millennium ago. Much of this region was commonly known as "Turkestan," or more appropriately, "Greater Persia." (Afghanistan was not even partitioned from what is today Pakistan until 1898.) Yet, given these rich mixtures of identities and groupings in Central Asia, and despite their short history as identifiable entities with definable borders, it remains a serious mistake to clump together all the states and peoples of the region into an identity simply called "the Stans."

2) Styles of leadership, if not the leaders themselves, are unlikely to change.

Much has been said about the Soviet-style leadership that has remained very much in power since independence. Saparmurat Niyazov, or better known as Türkmenbashi ("Father of the Turkmens"), considers himself a modern Atatürk, but is anything but that; despite widespread corruption and precipitously declining standards of living for his people, he survived November 2002 coup and assassination attempts, allegedly masterminded by a former top Turkmen official, Boris Shikhmuradov.[18] Perhaps more than anywhere else in Central Asia, the façade of democracy and the cult of personality are obvious and present in Turkmenistan.

In Uzbekistan, President Islam Karimov has deftly maneuvered himself and his state into a position as an ally with the United States in the war against terrorism, to include the establishment of an air base, Khanabad, in the south. At the same time, Karimov has received criticism for alleged human rights abuses, unwillingness to open the Uzbek Som to foreign exchange convertibility (disregarding the International Monetary Fund's "Staff Monitored Program"), and has repressed the practice of Islam to such an extent that much of the political opposition has been funnelled into support for radical terrorist groups, especially the Islamic Movement of Uzbekistan (IMU). By 2004, continued support for Karimov, from either Russia or the United States, seems unclear.

Kazakhstan's Nursultan Nazarbaev, former Communist party chief for the Soviet republic, has held sway over the state with a firm hand ever since independence; although allowing the appearance of democracy in Kazakhstan, he has regularly squelched any political attempts to mount an effective, and democratic, opposition movement. Equally, in Kyrgyzstan, former physicist (fond of quoting Thomas Jefferson and Kyrgyz poetry) Askar Akaev was once considered the most reformed and "liberal" leader in Central Asia; in recent years, his government has increasingly moved toward central, firm, and often repressive control. While Akaev has promised to step down by 2005, there is no clear mechanism or apparent intent to provide for the orderly transition of power in Kyrgyzstan or *anywhere* in Central Asian states (or in the Caucasus, for that matter). The one politically improved prospect in the region, to the surprise of most, is Tajikistan. With its economy virtually destroyed by a brutal civil war from 1992 to 1997, President Imomali Rakhmonov appears to have introduced the making of a coalition government promised at the end of the war. This coalition incorporates the influences of revivalist Islamic leaders and former Communist elites.[19] Tajikistan's economy, nonetheless, remains a disaster, with the 2003 state budget submitted at one-tenth the level of the 1990 budget.[20]

3) Borders won't be changing.

Despite the inconsistencies of leadership and the bleeding of nations across state territories, it seems unlikely that any of the borders of any state will likely change at any near time in the future. Despite their sometimes antagonistic relationship with each other, the five Central Asian states are still struggling with the consequences and costs of political and social transition.

4) Resources will demand the most "external" attention yet "foreign" influence—including U.S.—will not be as great as many believe.

Martha Brill Olcott, a Senior Associate at the Carnegie Endowment for International Peace and a long-time prudent observer of Central Asia, has suggested that the attraction of resources such as oil or natural gas, while evident to all in a geopolitical context, may not be transforming the region as radically as some suggest.[21] While true that the United States now has a base in Uzbekistan and an air base outside the capital of Bishkek, Kyrgyzstan, it is also true that Russia has intentionally moved to improve relations with its Kazakh, Uzbek, and Kyrgyz neighbors. In alleged support of a collective security treaty between these states, Russia also established its first-ever "overseas" base in late 2003 outside Kant, Kyrgyzstan.[22] Although modest in size (with ten aircraft and 300 troops), the base establishment is significant.

What does not appear to be happening is a conflictual relationship between the United States and Russia in Central Asia. Nominally, Russia continues to support the United States in its war on terrorism, and there does not appear to be a grand struggle such as the "Great Game" of the nineteenth century in the region. At the same time, there is little evidence that a strategic reorientation has suddenly taken place in Central Asia toward the United States. Indeed, despite the ongoing difficulties of Afghanistan, the United States appears to be focusing elsewhere in the world; inevitably then, the significance of Central Asian states (to the West, at least) may wane as well. For the region itself, dealing with each other and dealing with internal concerns will remain primary and ongoing challenges.

5) The race is on between economic distribution and human security over the next decade.

Kazakhstan, despite being what Olcott has termed an "unfulfilled promise," remains the only state in Central Asia to actually support positive economic growth since independence. Resources, and prospects for prosperity, therefore are the crucial challenges. Water is a critical challenge in this ongoing and uncertain struggle.

From a strategic perspective, the Caspian basin (located both in the Transcaucasians and Central Asia) is a tempting future energy source. With potential reserves of 160 billion barrels of oil (at an estimated current value of $4 trillion) and the world's largest natural gas deposits, the region is an acknowledged resource-rich environment. Indeed, many American-owned international oil companies have frequently criticized U.S. foreign policy, particularly sanctions against Iran, as damaging future U.S. economic interests in Central Asia in particular. Yet the region is also torn by ethnic and civil conflicts, some that have been going on for decades. Further, social security and standards of living for individual citizens have plummeted in the region since the end of the Soviet empire. According to Olcott, as well as other observers, Central Asian states are not plunging into chaos yet but the possibility still exists. Each of these states faces difficult choices in the first decade of the twenty-first century; according to even the most optimistic estimates, economic benefits from oil and gas resources will not be realized until 2010.

2. DEFINING ENVIRONMENTAL SECURITY

Moving from the specific aspects of geographic and geopolitical challenges in Central Asia, it may be appropriate to consider how theoretical approaches to "non-traditional" (or non-military) security issues have been considered elsewhere in the recent past. To do this correctly requires recognizing some fundamental shifts in security definition—particularly with the emergence of human and environmental security themes in the 1990s.

In terms of precise categorization, there are critical differences between human and environmental security. In the broadest sense, environmental security considers issues of environmental degradation, deprivation, and resource scarcity; by contrast, human security examines the impact of systems and processes on the individual, while recognizing basic concerns for human life and valuing human dignity. Yet as numerous examples illustrate, complex interactions within various environments often place stress on the security of the individual. Thus, environmental and human security often co-exists in a complicated interdependence best conceptually considered as "extended security."

Policy makers would be wise to recognize this conceptual approach.[23] Yet for research to be relevant to policy makers, it should almost always contextualize significance within a specific human- and regional-oriented perspective. To be blunt, there is a specific and pragmatic reason for emphasizing these issues in terms of security: doing so makes the topic both accessible for decision makers and provides a basis for determining present and future policy. Ole Wæver, one of the earlier influences (along with Barry Buzan) in promulgating the "new security agenda," reflects a certain skepticism, nonetheless, about the ability to influence policy through reframing the understanding of security:

> A security issue demands urgent treatment: it is treated in terms of threat/ defence, where the threat is external to ourselves and the defence often a technical fix . . . traditionally the state gets a strong say when something is about security. To turn new issues (such as the environment) into "security" issues might therefore mean a short time gain of attention, but comes at a long-term price of less democracy, more technocracy, more state and a metaphorical militarisation [sic] of issues. For this reason, environmental activists and not least environmental intellectuals who originally were attracted to the idea of "environmental security" have largely stepped down. . . . Security is about survival. . . . The invocation of security has been the key to legitimizing the use of force, and more generally opening the way for the state to mobilize or to take special force. . . . Security is the move that takes politics beyond the established rules of the game."[24]

There is, however, any number of overextending assumptions in the above reference. Above all is the assumption that security is an extreme term that can only be couched in terms of threat, and that the state—as political monolith— can only respond to with the use of force. Security is far richer in contextual meaning than such a stratified identity seems to allow.

Security is a basis for both policy *response* and long-term *planning*. Further, the use of force—particularly military force—is often an ineffectual and irrelevant response to the "new security agenda." Thus, the argument that "environmental security" is simply a mask for military intervention is argument that is, at best, thin. What *is* true is that the understandings of, and definitions for, environmental security range so broadly that its meaning takes on something for everyone—and perhaps, ultimately, nothing for no one.

For the specific relevance of the term "environmental security" applied to the significance of water resources in Central Asia, the broadest relevant definition should be, and should remain, an understanding that environmental security centers on a focus that seeks the best effective response to changing environmental conditions that have the potential to reduce stability, affect peaceful relationships, and—if left unchecked—could contribute to the outbreak of conflict. Perhaps ironically, the best overall definition for environmental security relevant to the Central Asian space was written two decade ago, when Norman Myers argued that

> National security is not just about fighting forces and weaponry. It relates to watersheds, croplands, forests, genetic resources, climate and other factors that rarely figure in the minds of military experts and political leaders, but increasingly deserve, in their collectivity, to rank alongside military approaches as crucial in a nation's security.[25]

In contrast to Wæver's suspicious pessimism regarding the true political motives for the environmental security agenda, and in support of Myers' above ideas on the need to rethink—and re-conceptualize—security, one would hope that both military and political leaders will come to widely recognize the validity of environmental security for strategy and policy initiatives. Based on the tensions, disagreements, and uncertainties regarding water in Central Asia, however, it is not clear—and, sadly, not likely—that this recognition will forestall some potentially disastrous outcomes in the region.

3. THREAT AND VULNERABILITY: THE DIFFERENCE BETWEEN THEM AND WHAT IT MEANS FOR CENTRAL ASIA

In the definitions offered above, much of the distinction centered on the best approach to dealing with environmental *threats*. Even those, such as Wæver, who first promulgated the idea of environmental security and then backed away, seemed to have done so because of the implication that security contextualization must be couched solely in terms of threat response with use of (almost always, military) force. Few of the definitions, with the possible exception of Myers, recognize that *vulnerability* can also be a key feature of the security calculus. Although few policy makers might immediately recognize the difference between the terms, both concepts suggest different realities.

A threat is identifiable, often immediate, and requires an understandable response. Military force, for example, has traditionally been sized against threats: to defend a state against external aggression, to protect vital national interests, and enhance state security. A threat, then, is either clearly visible or commonly acknowledged.

A vulnerability is often only an indicator, often not clearly identifiable, often linked to a complex interdependence among related issues, and does not always suggest a correct or even adequate response. While disease, hunger, unemployment, crime, social conflict, terrorism, narco-trafficking, political repression, and environmental hazards are at least somewhat related issues and do impact the security of states and individuals, the best response to these related issues, in terms of security, is not at all clear. Further, a vulnerability (unlike a threat) is not clearly perceived, often not well understood, and almost always a source of contention among conflicting views.

Thus, the *time* element in the perception of vulnerability can also further confound the problem—and make vulnerabilities far more controversial and far less pressing than the clear and present dangers of threats. Extreme vulnerability can arise from living under conditions of severe economic depravation, to victims of natural disasters, and to those who are caught in the midst of war and internal conflicts. Long-term human *development* attempts thus make little to no sense and offer no direct help. The situation, to be blunt, is not one of sustainability but of rescue.

But there are also cases of long-term vulnerability in which the best response is uncertain. Given this uncertainty, the frequent—and classic—mistake of the decision maker is to respond with the "gut reaction." Thus, the intuitive response to situations of clear ambiguity is, classically, to *do nothing at all.* The more appropriate response is to take an adaptive posture; to avoid the impulse to act purely on instinct; and to recognize what variables, indicators, and analogies from past examples might best inform the basis of action. Yet environmental and human security, since they are contentious issues, often fall victim to the *do nothing* response because of their vulnerability-based conditions in which the clearly identifiable cause and the desired prevented effect are often ambiguous.

What are the implications for transboundary water resources in Central Asia? Essentially, a number of vulnerability issues, if left unchecked over time, can take on significance that could easily match the challenges of the ongoing war on terrorism, can impact effective governance, and potentially can lead to conflict. To be specific, vulnerabilities, if left unchecked over time, *become* threats.

4. REVIEW OF ENVIRONMENTAL SECURITY AND STABILITY MODELS

A review of recent research suggests that some theoretical models have attempted, at least, to be all encompassing in their explanation for environmental performance as well as offering causality for potential conflict and stability impact. This brief review will consider some of these predictive models, to include interdependent factors that are "trigger mechanisms" that can unleash violent conflict, create socio-economic disparity, and induce long-term insecurity.[26] The application of these particular arguments for geo-specific research— such as water and other related (and interdependent) security issues in Central Asia effect one another—has not yet reached a point of precise refinement.

4.1 Environmental Performance and Human Security Indicators

Research conducted at the U.S. Naval War College, which bounded its own focus to the extent of the European Command's regional area of responsibilities— to include much of MENA (Mediterranean, Europe, and North Africa)— provides a sound basis from which to consider factors relevant to human security, environmental change, and regional application.[27] As S. R. Hearne argues, understanding and extrapolating from environmental indicators can prove effective in providing both feedback on, and assessing progress made, toward:

- Reporting on the state of the environment per national law or other agreements;
- Raising environmental issues onto the political agenda to promote further debate;
- Supporting policy development to address priority environmental concerns;
- Supporting efforts to address environmental problems during budget formulation;
- Measuring environmental performance and the success of policy responses;
- Identifying trends by major sectors, such as energy, agriculture, transport, industry;
- Establishing environmental targets at the sectoral and sub-national levels;
- Providing early warning to prevent environmental damage;
- Measuring progress toward sustainable development;
- Facilitating national, regional, and international environmental planning;
- Prioritizing regional intervention and engagement activities; and
- Communicating progress to the public and national and international institutions.[28]

Direct and indirect pressures therefore affect sustainable development over the long term and can potentially induce human security crises. Both the OECD and the European Union have developed frameworks intended to illustrate how states, and communities, might best respond to events and to develop alternative policies or behavioral responses.[29] Specifically, the OECD "Pressure State Response Model" (or PSR) framework illustrates in simple context the often complex causal relationships induced by environmental change and human societal response to or effect from such change.[30] It seems also worth noting that the World Bank in its own assessments has largely accepted the PSR framework in linking environmental problems to developing program objectives.[31]

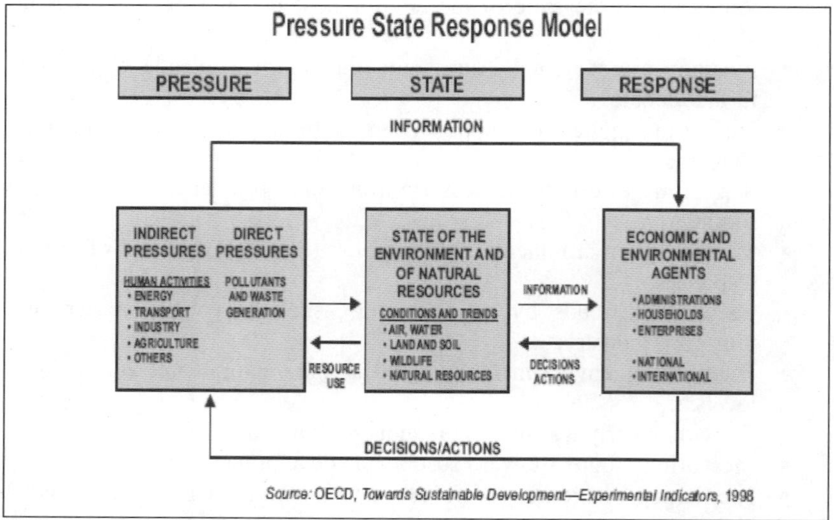

Figure 2: Pressure State Response Model

4.2 Sustainability Indices

Moving from an overall descriptive assessment of the nature of environmental driving forces and pressures on particular states, there have been a number of arguments for a single index of sustainable development from which to best assess and respond to situations of human security distress and environmental change. Perhaps the most well known of these indices is the United Nations Human Development Index (HDI), which correlates three specific factors that suggest the normative dimension of social sustainability—life expectancy, education, and Gross Domestic Product (GDP). While acknowledging that the widely referenced HDI is itself perhaps insensitive to some aspects of human security, the United Nations' own Commission on Sustainable Development has noted that there may indeed need to be some caution exercised regarding use of the HDI as an exclusive influence for national policy making.[32]

As an alternative index, one of the more intriguing frameworks that seeks to encompass a richer display of factors and influences than the HDI is the European Commission's Joint Research Centre example of the "Dashboard of Sustainability." So named because it is meant to metaphorically represent the clusters of indicators displayed on the dials and gages of a complex instrumentation panel (or a car dashboard), the framework means to monitor environmental quality, provide system feedback, reflect economic performance, and assess institutional factors.[33]

Notably, The Commission on Sustainable Development criticized the original "Dashboard" framework as simplistic and basic, and lacking insufficient detail to offer true merit as a policy tool.[34] A subsequent revision of this framework, nonetheless, appears to have incorporated this criticism, as the following example for the nation of Georgia attempts to demonstrate:[35]

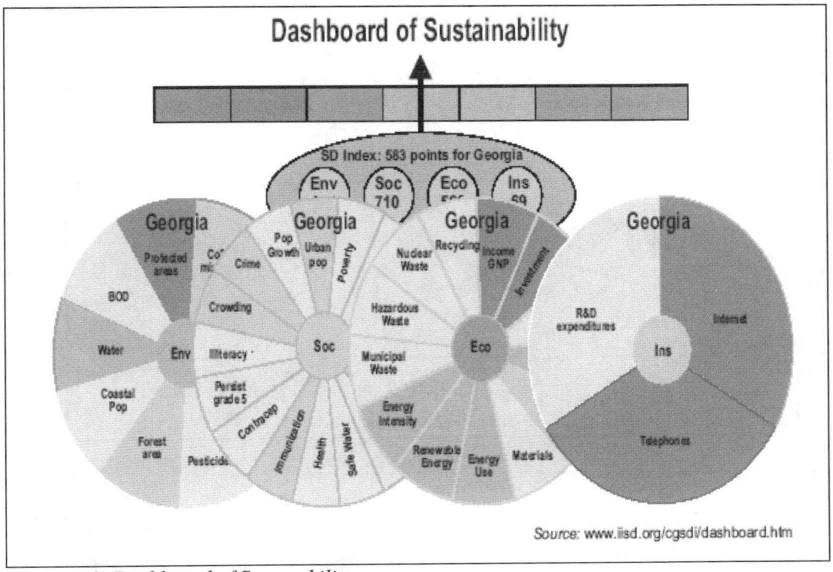

Figure 3: Dashboard of Sustanability

Similar indices are available for over one hundred nations. Performance indicators rely on a seven-color-coded system—dark red for worst outcome, dark green for best case—and specific policy performance in each dimension is also scored using a point system ranging from 0 (dark red) to 1000 (dark green). Finally, the software package calculates an overall Sustainable Development Index (SDI) for each country. Thus, while the "Sustainability Dashboard" may not represent the most widely accepted index, it does offer intriguing possibilities for geo-specific comparisons for states and communities.

An alternative sustainability index is the Environmental Sustainability Index (ESI). A collaborative effort between the World Economic Forum's Global Leaders for Tomorrow (GLT) Environmental Task Force, the Yale Center for Environmental Law and Policy (YCELP), and the Columbia University Center for International Earth Science Information Network (CIESIN).[36] First released at the World Economic Forum's annual meeting in 2001, the ESI ranked 122 countries according to their specific achievements in environmental sustainability—defined as the ability to produce significant results in five core component areas: environmental systems, reducing environmental stresses, reducing human vulnerability, social and institutional capacity, and global stewardship. An illustrative example of these five component factors is portrayed below, and attempts to articulate how specific states are measured for environmental sustainability.

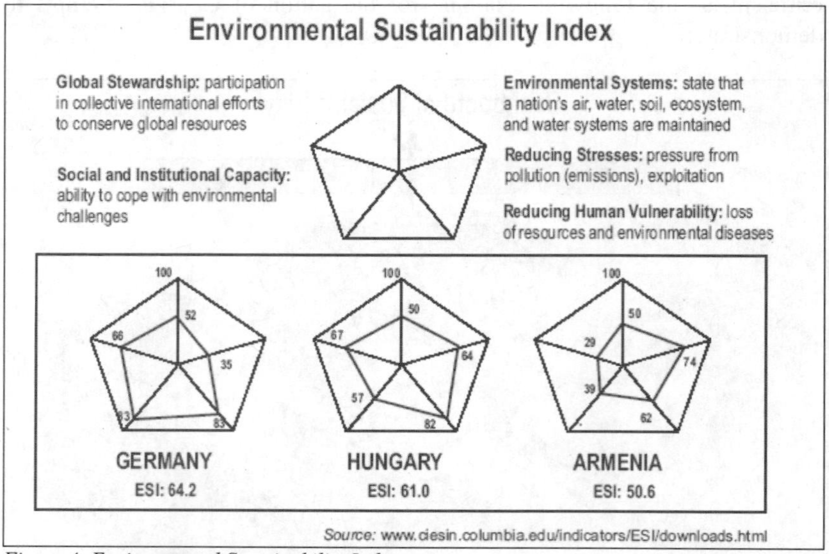

Figure 4: Environmental Sustainability Index

While some have also criticized the validity of the Environment Sustainability Index (in which, for example, the Russian Federation was ranked 33rd while Singapore is ranked overall 65th), the ESI research team has argued that the index serves as an underpinning tool that can form a viable "watch list of countries facing potential environment-driven crises."[37]

4.3 Stability Assessment Frameworks and Causal Explanations for Conflict

The concept of state failure became a dominant theme in the 1990s—partially as a result of the journalist Robert Kaplan's piece titled *The Coming Anarchy* (which appeared in the February 1994 issue of *Atlantic Monthly)* as well as with work that began at the Peace and Conflict Studies Program at the University of Toronto. The director of that program, Thomas Homer-Dixon, refined his earlier ideas in a 1999 book titled *Environment, Scarcity, and Violence.*[38] Drawing on more developed analysis than in previous research results, Homer-Dixon examined the causal links between socio-economic, political, and environmental "stressors" and proposes a "Core Model" of causal linkages. The resulting causal linkage model included feedback loops and provided for different stages for external (and internal) intervention. In contrast to earlier case studies on environmental scarcity, state failure, and conflict, later research recognized that to always assert that environmental scarcity as *specific* explanation for conflict is difficult, at best, to prove—since scarcity itself is "always enmeshed in a web of social, political, and economic factors."[39]

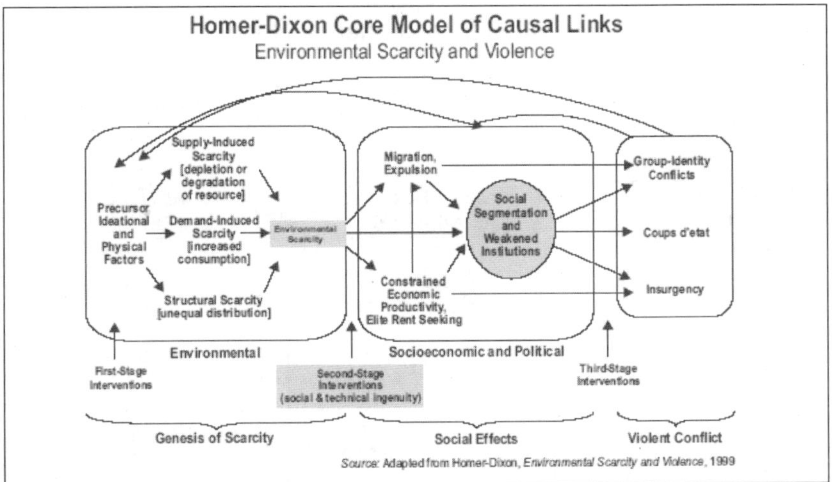

Note: The term "environmental scarcity," used in the Core Model reflects the scarcity of renewable resources, and accounts for supply-induced scarcity, demand-induced scarcity, and structural scarcity.[40] What is notable about this model's relevance to Central Asia is that—as concerns transboundary water resources—supply-induced scarcity, demand-induced scarcity, and structural scarcity are all occurring simultaneously.

Figure 5. Homer-Dixon Core Model of Causal Links

In developing this causal linkage model above, the Toronto research team employed a method of "process tracing": a step-by-step analysis of causal processes drawn from specific case studies, attempting to focus on "if" and "how" environmental scarcity contributes to violent conflict. Homer-Dixon's work suggests that environmental scarcity is mainly an indirect cause of violent conflict.

A second influential framework—which, intriguingly, was created *after* the publication of Kaplan's "Coming Anarchy" essay, and was meant as a partial response to the pessimistic assessments in Kaplan's argument—is the State Failure Task Force Phase II "Mediated Environmental Model." Claiming to be the first reported empirical large-scale study to investigate the critical factors most responsible for state collapse and failure, the State Failure Task Force was created in 1994 following a request by then–Vice President Al Gore.[41] At that time, there was a sense of increasing instability and collapse of governance in many nations of the world following the end of the Cold War.

Policy makers hoped, therefore, that research into state failure might provide indicators of early warning to facilitate suitable forms of international intervention. In response, the Central Intelligence Agency established the State Failure Task Force, to conduct a comprehensive examination of why certain states succeeded, while others seemed to fail. Research was conducted in a series of phases; for the purposes of this chapter, only results from the Phase II report appear.

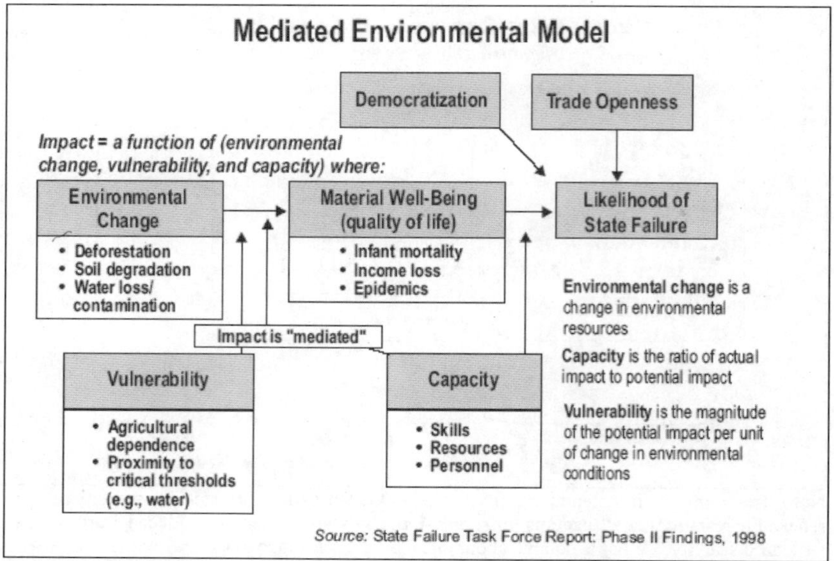

Figure 6. Mediated Environmental Model

The task force's predictive "Mediated Environmental Model" considered aspects of democratization, trade openness, environmental stress, material well-being, vulnerability, and capacity as they contribute to the likelihood of state failure. Equally, a separate model based on Sub-Saharan Africa results revealed the not surprising though disturbing indicator that partial democracies (as most of the states of Central Asia would prove at best to be) are at a relative risk of failure *eleven times greater* than an autocratic state under similar conditions of stress. Further, counter-intuitive issues such as "infant mortality" rates took on great significance in the study, because of their broader impact on other well-being issues. Such counter-intuitive issues, therefore, also constitute initial warnings of serious systemic problems.

While the aim of the argument here is *not* to suggest that state failure is imminent in Central Asia, there are some parallels to be drawn. Specifically, water is part of a larger complexity linked to security. Water, as a scarcity or poorly managed resource, can aggravate state governance with little or no interest in democracy and can also "fuel" opposition that in its extreme viably supports radical terrorism such as the IMU in Uzbekistan.

Major findings from the Phase II State Failure research that have some significance for Central Asia include:

- Partial democracies are particularly vulnerable and are at elevated risk of state failure;
- Gradual transition to democracy will likely improve the chances for success; and

- Ethnic discrimination alone may not be the most critical factor leading to conflict as was evident in a modified global model developed for Sub-Saharan Africa.

The State Failure Project Phase II research also investigated the impact of environmental change on material well-being as a function of national resource vulnerability and a state's institutional capacity to respond to the stressors associated with environmental change. Significant findings include:

- Environmental change does not appear directly linked to state failure; rather it is part of what has already been described as complex linkages and interaction among a number of socio-economic, political, and environmental stressors;
- The state's capacities to respond and by the degree its resources are vulnerable to environmental shock;
- Analyses are being hampered by a lack of long-range environmental data.[42]

This last assessment is clearly evident in the Central Asian dilemma today.

5. SUGGESTIONS FOR FUTURE RESEARCH

While acknowledging that such forecasts are *not* strict predictions—but rather *projections*—from reasonable indicators of current and recent trends and effects, there can be little doubt that significant change may well occur in the Central Asian context regarding transboundary water resource issues. Such change will directly affect the security calculus of the entire region. As such, a number of general conclusions and concerns could be raised about the shifting landscape of Central Asia and the critical uncertainties that will inevitably emerge:

1. Transboundary water resources form part of a larger network of cultural, political, and economic linkages in Central Asia. To date, little multilateral agreement has been reached that will solve the wider negative consequences of collapsing "extended security," exemplified by the potential impending water crisis in the region.
2. Unless significant agreements and institutional agreements are reached, there will equally be little incentive to establish or sustain early-warning and conflict prevention centers for Central Asia.
3. The specific relevance of the term "environmental security" to Central Asia should be framed as an understanding that environmental security centers on focus that seeks the best effective response to changing environmental conditions that have the potential to reduce stability, affect peaceful relationships, and—if left unchecked—could contribute to the outbreak of conflict.
4. A number of *uncertain vulnerabilities* must be vividly presented to policy makers as having serious long-term consequences, to include their eventual emergence as *threats*. These vulnerabilities in Central Asia may include— but are certainly not limited to—disease, hunger, unemployment, crime, social conflict, terrorism, narco-

trafficking, political repression, and environmental hazards. Vulnerabilities, if left unchecked over time, *become* threats. This is clearly relevant to the Central Asian space, where we may have moved from a dynamic of the old U.S.-USSR *security dilemma* to encompass issues in the twenty-first century that will also include a new *survival dilemma* for a vast percentage of the human population in the region. The forced movement of the residents of Karalkapakstan as a result of the ecological ruin of the Aral Sea may only be the first example.

5. Military and political leaders should recognize the validity of environmental security for strategy and policy initiatives. The application of these particular arguments for geo-specific research has *not* reached a point of precise refinement. Models, developed to date, remain insufficient.
6. Further research must concentrate on obtaining reliable and broad indicators of environmental change measured over longer periods of time. Although recent research suggests vastly increased upswings in ecological/natural disasters in Central Asia, more data is required.
7. We must recognize the relevance—and the danger—of adding the term "security" to either environmental or human-centered concerns. To be blunt, there is a specific and pragmatic reason for emphasizing these issues in terms of security: doing so makes the topic both accessible for decision makers and provides a basis for determining present and future policy.
8. Solving water issues in Central Asia may help solve socio-economic issues such as the transition to democratic, or stable, governance in each state and mitigating the causes and effects of transnational terrorism in the region.

Ultimately, the inconclusive and sometimes contradictory results of various models and frameworks relevant to an examination of "extended security" in Central Asia leave us in a state of uncertain certainty. Environmental change is occurring now that clearly will affect the security calculus, but we simply do not know enough or have available data for definitive proof. Further, while we may recognize vulnerabilities, or aspects of what may be vulnerabilities, we simply do not know which will emerge as threats.

Thus, researchers who might desire a specific quantitative outcome in studying and applying focus to geo-specific regions, such as the Central Asian space, are still unable to "bound" expected outcomes and prevent potential future negative security events. Such modeling could be inherently dangerous, however, particularly if used as the exclusive basis for foreign policy decisions to intervene or abstain in attempting to control the boundaries of a collapsing state or region. Models, driven by quantitative outcome answers, may ignore the value of qualitative process examinations that help frame more appropriate questions. Further, predictive modeling may lead as well to determining "permanent failure states": those states that, based on quantitative analysis alone, would appear to have no possibility of social, political, or economic recovery. One possible alternative to projective modeling, therefore, might include a more balanced focus on recognizing the complex interdependence of socioeconomic, environmental, and institutional factors. A recently developed framework that offers this balanced approach—simple in concept, immensely difficult in practice—is the "Stability Pyramid." [43]

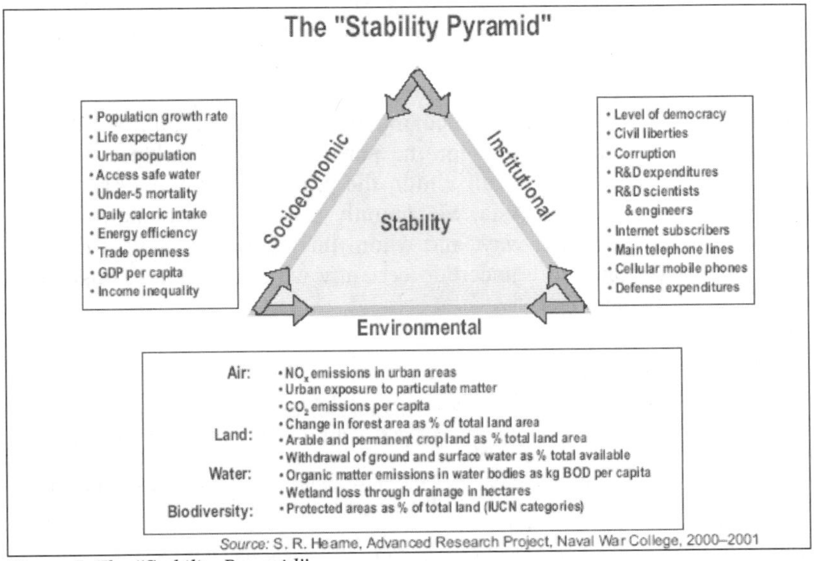

Figure 7. The "Stability Pyramid"

This "Stability Pyramid" may also point to a crucial missing link in the security dilemma that has characterized the 12 September world (as well as the post–Cold War world). While much discussion came about following the publication of Francis Fukuyama's essay "The End of History?" about the critical importance of democratization and (economic) liberalization, little relative weight was given to the systemic and structural importance of environmental aspects such as scarcity. Specifically, Fukuyama first argued in 1989 that, with the end of the Cold War, specific nations and regions may have reached "the end of history," in which Western liberal democracy will represent the final form of human government. Admittedly, a decade later, Fukuyama recognized that he did not sufficiently account for the social, cultural, political—and military—effects of globalization, information technology, and biotechnology in making his original argument. He insisted, however, "Nothing that has happened in world politics or the global economy in the past ten years that challenges, in my view, the conclusion that liberal democracy and market-oriented economies are the only viable options for modern societies."[44] Yet the crucial linkage of environment to security in some societies, such as in Central Asia, may only now be emerging as a recognition of equal importance and a new reality of pressing need.

Ultimately, the complexity of human interaction with and response to complex environmental influence may well lie beyond any viable or accurate modeling attempt. Admittedly, this is a contentious conclusion, but probably an honest one as well. Reasonable strategies for regional security and ecological stability in Central Asia thus need to balance preoccupation with rationality with the recognition that policy makers almost always lack clairvoyance, suffer from cultural blinders, and are often driven by contingency responses. These studies should further recognize reasonable alternative appraisals, the frequent lack of will and resources, and the nature of the overall desired goals of multiple players involved in any scenario. Most often, these crucial actors enter into a kind of psychological—and sometimes physical—St. Vitus' dance until exhaustion or resolution set in.

At best, we should hope for the participation of multiple decision makers and for the desire for all within the region to involve. And as we assess the future of Central Asia, we can only know that dynamic change is coming—but how, in what way, and whom this change will most directly affect remains yet one more uncertain certainty we must always consider in defining this critical geography.

NOTES

1 "Water Scarcity and Conflict," in *Security Challenges of the 21st Century* (Peter Lang: Bern/ Frankfurt: 1999), 11, http://www.padrigu.gu.se/ohlsson/files/Bonn.pdf (9 November 2003).

2 As reported by Agence France Presse. Available through the *New York Times* online, under the title "UN Forum Told One Billion Lives Threatened by Bad Water," http://www.nytimes.com (3 September 2003).

3 National Intelligence Council, *Global Trends 2015: A Dialogue about the Future with Nongovernmental Experts*, NIC Paper 2000–02 (December 2000). Available at: http://www.odci.gov (under "Publications"). See also Michael T. Klare's "The New Geography of Conflict," *Foreign Affairs*, May/June 2001, pp. 49–61, as well as his *Resource Wars: The New Geography of Global Conflict* (New York: Metropolitan Books, 2001).

4 As reported by Agence France Presse. Available through the *New York Times* online, under the title "Water Issues Erupt at Central Asia Forum," http://www.nytimes.com (3 September 2003). These events were separately confirmed by a UN delegate in private conversation with the author.

5 Not all water use is for agriculture, however. Certainly one of the more seemingly frivolous projects is Turkmenistan's diversion of water from the Amu Darya (before it flows into Uzbekistan) for the construction of a huge artificial lake in the Karakum Desert, 24 *Chapter 1* known as the Lake of the Golden Century. Turkmenistan claims the lake is intended only to collect drainage and runoff, but the entire project has ecological risks—to include turning drainage areas into swampland. Although the project itself was in the planning stages in the 1940s under Stalin, work did not actually begin until October 2000. Some also speculate that a nuclear power station may be built in the area, which would use the lake waters for cooling purposes.

6 The Syr Darya flows from northern Kyrgyzstan and through Uzbekistan before crossing into Tajikistan; it subsequently re-enters Uzbek territory before flowing through southern Kazakhstan, ending course in the northern Aral Sea basin. The Amu Darya flows from the Tajik-Afghan border along the Uzbek-Afghan border (whose territories the river demarcates); it continues into Turkmenistan, flows along the Turkmen-Uzbek border before crossing into Uzbekistan, ending course in the northern Aral Sea basin.

7 *Report to the 16th Plenary Sessions of the Uzbekistan Communist Party Central Committee*, quoted in Iwao Kobori and Michael H. Glantz (eds.), *Central Eurasian Water Crisis: Caspian, Aral and Dead Seas* (United Nations University), 1998, referenced in the International Crisis Group Asia Report N°34, *Central Asia: Water and Conflict,* 6.
8 Charles William Maynes, "America Discovers Central Asia," *Foreign Affairs*, March/ April 2003. Access through ProQuest Direct.
9 Interview, 20 February 2002, International Crisis Group Asia Report N°34, *Central Asia: Water and Conflict,* 7.
10 *Soglashenie o sotrudnichestve v upravlenii ispolzovania i okhrany vodnykh resurvov iz mezhgosudarstvennykh istochnikov* [Agreement on Coöperation in the Management of the Use and Protection of Water Resources from Interstate Resources], reported in *Central Asia: Water and Conflict*, 7.
11 International Crisis Group interviews with Yusup Kamalov, Chairman of the Union for the Defence of the Aral Sea and the Amu Darya, Nukus, Karalkapakstan, 23 January 2002, as well as Emazar Makhmudov, Director, Institute of Water Problems of the Uzbek Academy of Sciences, Tashkent, 21 January 2002.
12 Adapted from Ohlsson, 1, http://www.padrigu.gu.se/ohlsson/files/Bonn.pdf (9 November 2003).
13 Ibid., 2; 11–15.
14 The term "morbid interregnum" is taken from the Italian Communist Antonio Gramsci, who in the 1920s declared: "The old order has died. The new is not yet born. And in the interregnum, the morbid symptoms appear."
15 As one example of how this recognition has affected other regional strategies that might influence decision making, see *The Barcelona Process: The Euro-Mediterranean Partnership, 2001 Review,* 2nd edition (Brussels: European Commission, 2002), 13, available through the Europa server (http://europa.eu.int).
16 Rajan Menon, "The New Great Game in Central Asia," *Survival*, Volume 45, Number 2 (Summer 2003): 201.
17 From the Perry-Castañeda Library Map Collection at the UT [University of Texas, Austin] Library Online, http://www.lib.utexas.edu/maps/index.html. Specific reference is: http://www.lib.utexas.edu/maps/middle_east_and_asia/casia_ethnic_93.jpg.
18 During the author's work in Turkmenistan in 1996, numerous youth mentioned privately the theme that "No matter what you are told here, everything is a lie." Conditions do not seem to have changed since that time; indeed, to some analysts, Turkmenistan is regarded today as a kind of North Korea without nuclear weapons. *Critical Geography* 25
19 Notably, during the author's work in Tajikistan during the time of its civil war, he encountered a nuer of Tajiks on their way to or returning from fighting with the Northern Alliance against the Taliban in Afghanistan. To a man, each of the Tajiks the author spoke with considered life in Afghanistan far preferable to Tajikistan at that time.
20 International Crisis Group Asia Report N°51, *Tajikistan: A Roadmap for Development*, 24 April 2003, 2.
21 Martha Brill Olcott, "Taking Stock of Central Asia," *Journal of International Affairs*, Volume 56, Issue 2 (Spring 2003): 3–17. Martha Brill Olcott, "The Caspian's False Promise," *Foreign Policy*, Volume 111 (Summer 1998): 94–113.
22 Steven Lee Myers, "World Briefing | Europe: Russia: A Base Abroad," *New York Times*, 23 September 2003, Late Edition, A4: 8.
23 In furthering this argument, this chapter intentionally blurs the distinction between environmental and human security, in favor of an extended security approach. For the best overall writing and conceptualization of "extended security," reference the work of Emma Rothschild, Director of the Centre for History and Economics, Kings College, Cambridge University. As difficult as such "new" conceptions of security are, it seems worth noting that the United Nations Commission on Human Security continues to grapple with the definition of "security."
24 Quoted in Thomas Scheetz, "The Limits to 'Environmental Security' as a Role for the Armed Forces"; paper provided by the author. Wæver's original remarks, titled "Security Agendas Old and New, and How to Survive Them," were prepared for the workshop on "The Traditional and New Security Agenda: Influence for the Third World," Universidad

Torcuato Di Tella, Buenos Aires, Argentina, 11–12 September 2000.
25 Norman Myers, "The Environmental Dimension to Security Issues, *The Environmentalist*, 1986, 251.
26 Much of this section's ideas and influence originated with an advanced research project at the U.S. Naval War College during the academic year 2000–01. My thanks and acknowledgements, therefore, are due to Steven R. Hearne, an environmental specialist with the U.S. European Command, whose work stimulated this inquiry. Hearne's project was later published as *Environment Indicators: Regional Stability and Theater Engagement Planning*, AEPI-IFP-1001A (Atlanta, Georgia: U.S. Army Environmental Policy Institute, October 2001).
27 I should note that the European Command's AOR (Area of Responsibility) is actually quite broad, encompassing geographic focus on 55 countries in the Euro-Mediterranean and Sub-Saharan Africa. Further, although functional responsibility for the former Soviet republics of Central Asia has now shifted to the U.S. Central Command, all former Soviet republics in Central Asia are members of the Partnership for Peace (PfP) program and the Organization for Security and Coöperation in Europe (OSCE).
28 Hearne, 3–4.
29 For the purposes of the broad conceptual approach used in this chapter, the terms "framework" and "model" are used interchangeably. Some might clearly find this tact discomforting. As Hearne notes in his research, "The terms framework and model are often used interchangeably in different publications, [sic] however, framework implies a conceptual or basic arrangement or structure, whereas, a model is representative of some existing system. Robert Keen and James Spain define a model as "any representation of a real system [involving] words, diagrams, mathematical notation, or physical structures in representing the system . . . [the term] may have the same meaning as concept, hypothesis, or analog . . . [and] it must always involve varying degrees of simplification or abstraction." They 26 *Chapter 1* use the examples of the food chain and the ecosystem to illustrate a "conceptual model." See Robert E. Keen and James D. Spain, *Computer Simulation in Biology: A Basic Introduction* (John Wiley and Sons, New York, 1992), 2–3.
30 The DPSIR model was derived in some measure from, and is more compatible (than the PSR framework) with, the United Nations CSD [Commission on Sustainable Development] Driving Forces-State-Response (DSR) model. Jesinghaus, *A European System of Environmental Pressure Indices,* 2.
31 Lisa Segnestam, *Environmental Performance Indicators: A Second Edition Note*, Paper No. 71 (Washington, D.C.: World Bank, October 1999), 5–8.
32 United Nations Division for Sustainable Development, *Report on the Aggregation of Indicators of Sustainable Development*, Background Paper No. 2, for the Ninth Session of the Commission on Sustainable Development, 16–27 April 2001, New York, 5–21, http://www.un.org/esa/sustdev/csd9/csd9_docs.htm (5 April 2001).
33 The Sustainability Dashboard was developed in coördination with the International Institute for Sustainability's Consultative Group on Sustainable Development Indices. Peter Hardi and Alan AtKisson, "The Dashboard of Sustainability," Draft Design Specifications Document for the Consultative Group on Sustainable Development Indicators October 1999, 3–4, http://www.iisd.org/cgsdi/dashboard.htm (3 April 2001).
34 United Nations Division for Sustainable Development, *Report on the Aggregation of Indicators of Sustainable Development*, 117–18.
35 Joint Research Centre, "The CGSDI Dashboard of Sustainability—Version 3.3," 1, http://esl.jrc.it/envind/dashbrds.htm (16 March 2001). See also: Joint Research Centre, "The Methodology used for the Dashboard Software Tool," http://esl.jrc.it/envind/db_meths.htm (16 March 2001), 1–3.
36 Global Leaders for Tomorrow Environmental Task Force, World Economic Forum, 2001 *Environmental Sustainability Index*, Report to Annual Meeting (Davos, Switzerland, 2001), http://www.ciesin.columbia.edu/indicators/ESI/downloads.html).
37 *Environmental Sustainability Index*, 15.
38 Thomas Homer-Dixon, *Environment, Scarcity, and Violence* (Princeton, N.J.: Princeton University Press, 1999).
39 Val Percival and Thomas Homer-Dixon, "Environmental Scarcity and Violent Conflict: The Case of South Africa," *Journal of Peace Research*, Number 3 (May 1998): 294–295.

40 *Environment, Scarcity, and Violence,* 98; 134.
41 Daniel C. Esty, Jack A. Goldstone, Ted Robert Gurr, Barbara Harff, Marc Levy, Geoffrey D. Dabelko, Pamela T. Surko, and Alan N. Unger, *State Failure Task Force Report: Phase II Findings*, Environmental Change and Security Project Report (Issue 5, Summer 1999): 49.
42 Ibid., 24–26.
43 Hearne, 50. This framework was originally developed as part of an advanced research project at the U.S. Naval War College and later published in *Environment Indicators: Regional Stability and Theater Engagement Planning*, AEPI-IFP-1001A (Atlanta, Georgia: U.S. Army Environmental Policy Institute, October 2001).
44 See, in particular, Francis Fukuyama, "The End of History?" *National Interest*, Summer 1989, 3–18, and "Second Thoughts: The Last Man in a Bottle," *National Interest*, Summer 1999, pp. 16–33. *Critical Geography* 27

LIST OF WORKSHOP PARTICIPANTS

Directors:

Prof. Hartmut VOGTMANN
President of the Federal Agency for Nature Conservation,
Konstantinstr. 110, 53173 Bonn,
Phone: +49 228 8491 210
Fax: +49 228 8491 250
Email :
Hartmut.Vogtmann@BFN.de
GERMANY

Prof. Nikolai DOBRETSOV
President of the Siberian Branch of Russian Academy of Sciences,
prosp. Akademika Lavrentieva, 17,
Novosibirsk, 630090
Phone: +007 (3832) 30 05 67
Email: dobr@sbras.nsc.ru
RUSSIA

Lecturers or Key Speakers

John Anthony ALLAN
School of Oriental and African Studies, University of London, England;
Address: Thornhaugh Street, Russel Square, London WC1H0XG;
Fax: +44 207 229 3215
Email: tony.allan@kcl.ac.uk , tal@soas.ac.uk
UK

Joachim BENDOW
Executive Secretary of the International Commission for the Protection of the Danube River (ICPDR), ICPDR Pf 500, VIC, Room Do 443 A- 1400, Vienna;
Fax: + 43-1-26060 5895
Email: ispdr@unvienna.org
AUSTRIA

Mikhail BOLGOV
Water Problems Institute Russian Academy of Sciences, Head of the Caspian Sea laboratory, Russia
Tel/Fax: (095) 951- 68-41
Fax: (095) 135- 54- 15
Email: Caspian-sea@mtu-net.ru
RUSSIA

Jean-Francois CRETAUX
Research Engineer, CNES/LEGOS, France
Address: 18 Avenue Edouard Belin, 31400 Toulouse, France
Fax: + 33-561 253 205
Email :
JeanFrancois.Cretaux@notos.cst.cnes.fr
FRANCE

Shlomi DINAR
John Hopkins University, School of Advanced International Studies, USA
Address: 1740 Massachusetts Avenue, N.W. Washington, D.C. 20036 – 1983
Fax: 202- 663- 5769
Email: sdina@sais-jhu.edu
USA

Feliks EINHORN
Scientific Information Center of Interstate Coordination Water Commssion (SIC ICWC)
Address: 11 Building, Karasu –4, Tashkent, 700187, Uzbekistan
Phone: (998 71) 166 50 95, 166 50 96
Fax: (998 71) 166- 50- 97
UZBEKISTAN

Nikolai GRISHIN
Agency for Environmental Assessments «ECOTERRA», Russia
Address: P.O. Box 100, Moscow. 123423
Phone: (095) 108 8324
Email: ngrishin@online.ru
RUSSIA

Petra HELLEGERS
Researcher, Agricultural Economics Research Institute, Den Haag, Netherlands
Address: P.O. Box 29703 2502 LS The Hague
Fax: + 31 – 317 482745,
Email: Petra.Hellegers@wur.nl
NETHERLANDS

Heide JEKEL
Senior civil servant, Federal Ministry for the Environment, Nature Conservation and Nuclear Safety, Division Water Law, Germany
Address: Robert-Schumann-Platz 3D – 53175, Bonn
Fax: + 49- 1888 305 2693;
Email: heide.jekel@bmu.bund.de
GERMANY

Ragnar KINZELBACH
Dep. Of Biodiversity Research, University of Rostock, Germany;
Address: Universitaetsplatz 2, 18055 Rostock;
Fax: + 49 –381 498 6262
Email: ragnar.kinzelbach@biologie.uni-rostock.de
GERMANY

Peter LIOTTA
US Navy War College, USA
Address: 6686 Cushing Road Newport, RhodeIsland 02841
Fax: + 1 401 841 3893
Email: liottap@nwc.navy.mil
USA

Igor MALKOVSKY
Deputy director of Institute of Geography, Kazakhstan
Address: 99, Pushkina Str., Almaty, 480100
Phone (73272) 918129, 918091
Fax: (73272) 918102
Email: ingeo@mail.kz
KAZAKHSTAN

Ramiz MAMEDOV
Corresponding – Member of Azerbaijan Academy of Sciences, Institute of Geography, Azerbaijan Academy of Sciences ; Azerbaijan.
Address: 31, Gusein Jzavid St., Baku, 370143
Phone/Fax: (994 12) 39 35 41
Email: ramiz.mamedov@geo.ab.az
AZERBAIJAN

Dushen MAMATKANOV
Director of Institute of Water Problems and Hydropower, Bishkek, Kyrgyzstan
Phone: (996 312) 214572
Fax: .(996 312) 210674
Email: iwp@istc.kg
KYRGYZSTAN

Murat MUKHAMEDZHANOV
General Secretary for Scientific Affairs of National Academy of Sciences of Republic of Kazakhstan
Address: 28, Shevchenko Str., Almaty, 480021
Phone: (7 3272) 69 55 31,
Fax: (7 3272) 62 58 00
Email: m.mukhamedzhanov@nas.kz
KAZAKHSTAN

Inom NORMATOV
Director of Institute of Water Problems, Hydropower and Ecology, Academy of Sciences, republic of Tajikistan
Address: 33, Rudaki Ave., 734025, Dushanbe, Tajikistan
Email: owp@tojikiston.com
TAJIKISTAN

Aleksey PTITSYN
Director of Institute of Natural Resources SB RAS, Chita, Russia
Address: 16, Nedorezova St., Chita, 672014;
Phone: (3022) 211 689
Fax: (3022) 212 582
Email: root@cinr.chita.su
RUSSIA

Shavkat RAKHIMOV
Institute "SANIIRI", Uzbekistan
Phone: (10998712) 651 853, (10998712) 658 504
Email: saniiri@tkt.uz
UZBEKISTAN

Elena RODINA
Kyrgys- Russian Slavic University, Dept. of Sustainable Development of Environment; Kyrgyzstan
Address: 44 Kievskaya St., No. 3/ 113 apr. Bishkek, 720040
Phone: (9963) 431382; 282 776,
Fax: (996312) 286 492; 282776
Email: aral@krsu.edu.kg
KYRGYZSTAN

David SMITH
Economist, Division of Policy Development and Law, UNEP HQ, Kenya,
Address: P.O. Box 30552, Nairobi
Email: David.Smith@unep.org
KENYA

Oleg VASILIEV
Member of the Academy of Russian Academy of Sciences, Institute for Water and Environmental Problems (IWEP), Russia
Address: 2, Morskoy Prospect, Novosibirsk, 630090;
Phone/Fax: (3832) 302005;
Email: vasiliev@ad-sbras.nsc.ru
RUSSIA

Yuri VINOKUROV
Director of Institute for Water and Environmental Problems (IWEP), Russia
Address: 1, Molodezhnaya St., Barnaul, 656038
Phone: (3852) 367 855,
Fax: (3852) 240 396
Email: iwep@iwep.ab.ru
RUSSIA

Other Participants

Karl-Heinz ERDMANN
Head of Department II 1.2
Federal Agency for Nature Conservation
Address: Robert-Schumann-Platz 3, 53175 Bonn
Tel: +49 228 305 4466
Email: Erdmannk@BFN.de
GERMANY

Alain JUBIER
NATO HQ, Scientific Affairs Division, Belgium
Address: Boulevard Leopold III
B- 1110 Brussels
Phone: +32 02 707 4228
Fax: +32 02 707 4232
Email: a.jubier@hq.nato.int
BELGIUM

Jean KLERKX
President of International Bureau
for Environmental Studies,
Address: Rue Audrey Hepburn,
9/13, B 1090 Brussels
Fax: 07 3832 301062
Email: jean.klerkx@pandora.be
BELGIUM

Tatyana PAPINA
Dr., Institute for Water and
Environmental Problems (IWEP)
Address: 1, Molodezhnaya St.,
Barnaul, 656038
Phone: (3852) 358977
Fax: (3852) 240 396
Email: papina@iwep.ab.ru
RUSSIA

Heinrich SCHMAUDER
Senior civil servant, Department II
1.3,
Federal Agency for Nature
Conservation
Address: Konstantinstr. 110,
53179 Bonn
Phone: +49 228 8491 241
Fax: +49 228 8491 245
Email : Schmauderh@BFN.de
GERMANY

Arno TODT
Scientific Staff, Department II 1.2,
Federal Agency for Nature
Conservation,
GERMANY